ANIMAL AND PLANT *Anatomy*

VOLUME CONSULTANTS

• Barbara Abraham, *Hampton University, VA* • Amy-Jane Beer, *Natural history writer and consultant* • Allan Bornstein, *Southeast Missouri State University, MO* • John Friel, *Cornell University, NY* • Bill Kleindl, *Natural history writer and researcher* • Richard Mooi, *California Academy of Sciences, San Francisco, CA* • Ray Perrins, *Bristol University, England*

8

Respiratory system – Sequoia

Marshall Cavendish
Reference
New York

Contributors

Roger Avery; Richard Beatty; Amy-Jane Beer; Erica Bower; Trevor Day; Erin Dolan; Bridget Giles; Natalie Goldstein; Tim Harris; Christer Hogstrand; Rob Houston; John Jackson; Tom Jackson; James Martin; Chris Mattison; Katie Parsons; Ray Perrins; Kieran Pitts; Adrian Seymour; Steven Swaby; John Woodward.

Consultants

Barbara Abraham, Hampton University, VA; Glen Alm, University of Guelph, Ontario, Canada; Roger Avery, Bristol University, England; Amy-Jane Beer, University of London, England; Deborah Bodolus, East Stroudsburg University, PA; Allan Bornstein, Southeast Missouri State University, MO; Erica Bower, University of London, England; John Cline, University of Guelph, Ontario, Canada; Trevor Day, University of Bath, England; John Friel, Cornell University, NY; Valerius Geist, University of Calgary, Alberta, Canada; John Gittleman, University of Virginia, VA; Tom Jenner, Academia Británica Cuscatleca, El Salvador; Bill Kleindl, University of Washington, Seattle, WA; Thomas Kunz, Boston University, MA; Alan Leonard, Florida Institute of Technology, FL; Sally-Anne Mahoney, Bristol University, England; Chris Mattison; Andrew Methven, Eastern Illinois University, IL; Graham Mitchell, King's College, London, England; Richard Mooi, California Academy of Sciences, San Francisco, CA; Ray Perrins, Bristol University, England; Kieran Pitts, Bristol University, England; Adrian Seymour, Bristol University, England; David Spooner, University of Wisconsin, WI; John Stewart, Natural History Museum, London, England; Erik Terdal, Northeastern State University, Broken Arrow, OK; Phil Whitfield, King's College, University of London, England.

Marshall Cavendish
99 White Plains Road
Tarrytown, NY 10591–9001

www.marshallcavendish.us

Library of Congress Cataloging-in-Publication Data

Animal and plant anatomy.
p. cm.
ISBN-13: 978-0-7614-7662-7 (set: alk. paper)
ISBN-10: 0-7614-7662-8 (set: alk. paper)
ISBN-13: 978-0-7614-7672-6 (vol. 8)
ISBN-10: 0-7614-7672-5 (vol. 8)
1. Anatomy. 2. Plant anatomy. I. Marshall Cavendish Corporation. II. Title.

QL805.A55 2006
571.3--dc22

2005053193

Printed in China
09 08 07 06 1 2 3 4 5

Marshall Cavendish
Editor: Joyce Tavolacci
Editorial Director: Paul Bernabeo
Production Manager: Mike Esposito

The Brown Reference Group plc
Project Editor: Tim Harris
Deputy Editor: Paul Thompson
Subeditors: Jolyon Goddard, Amy-Jane Beer, Susan Watts
Designers: Bob Burroughs, Stefan Morris
Picture Researchers: Susy Forbes, Laila Torsun
Indexer: Kay Ollerenshaw
Illustrators: The Art Agency, Mick Loates, Michael Woods
Managing Editor: Bridget Giles

Photographic credits, volume 8

Cover illustration is a composite of anatomical illustrations found within this encyclopedia.

Corbis: Gehman, Raymond 1149; **Erica Bower:** 1050; **FLPA:** Dirscherl, R. 1103; Hosking, David 1148; Lewis, Linda 1097; Morrison, Reg/Auscape/Minden Pictures 1020; Wilson, D. P. 1129, 1130, 1135; Wu, Norbert 1105; **NHPA:** Bannister, Anthony 1075; Heuclin, Daniel 1078; Image Quest 1091, 1094; **Nature PL:** Allan, Doug 1113, 1114; Burton, Dan 1102; Cancalos, John 1083; Cole, Brandon 1067, 1068; Davidson, Bruce 1021; Gormesall, Chris 1140; McEwan, Duncan 1147; Munier, Vincent 1119; Petrinos, Constantinos 1029, 1098, 1107; Ruiz, Jose B. 1025; **OSF:** Packwood, Richard 1125; Seubert, Susan 1140; Soury, Gerard 1123; Wechsler, Doug 1133; **Photodisc:** 1041, 1085; 1136, **Photos.com:** 1013, 1014/1015, 1033, 1059; **SPL:** Douwma, Georgette 1127; Eye of Science 1030, 1056; **Still Pictures:** Bavendam, Fred 1107; Vezo, Tom 1057; Blakes, Rob 1139; Eichaker, X. 1064, 1065; Fairchild, Michael 1035; Fischer, Heike 1054; Gornacz, George 1100; Harvey, Martin 1047; Hodge, Walter H. 1131; Kage, Manfred 1024; Kazlowski, Steven 1117; Pfletschinger, Hans 1076; Seitre, Roland 1031; Tack, Jochen 1049; Wu, Norbert 1061, 1070, 1073, 1111.

Contents

Respiratory system

Energy is a requirement for life. Organisms get the energy they need through a process called respiration, which generally requires the presence of an essential gas—oxygen. Cells need oxygen to react with molecules from food, typically in the form of the sugar glucose. The reaction liberates energy; it also leads to the formation of molecules of water and another gas, carbon dioxide. This gas must be removed from the organism; the release of carbon dioxide and the simultaneous uptake of oxygen are called gas exchange. The liberation of energy using oxygen is called aerobic respiration. Without it, few of an organism's life processes could take place.

An anaerobic life

However, cells in tissues such as muscle sometimes use up oxygen faster than it can be supplied—for example, during intense exercise. Cells then switch to a different method of energy production called anaerobic respiration. Again, glucose is broken down to release energy, but with a chemical called lactic acid created as a by-product. Anaerobic respiration cannot continue for long; lactic acid builds up and must be removed. This buildup causes symptoms such as a stitch and muscle fatigue. However, anaerobic respiration can be crucial for providing a sudden burst of speed. This ability is so important that large sections of an animal's body can be devoted to providing for these occasional anaerobic bursts. Fish often have small bundles of red aerobic muscle running along their body. This is used for slow swimming. Vast banks of white anaerobic muscle flank the red muscle; they kick into action when the fish pursues prey or avoids a predator.

Why bother with oxygen?

Anaerobic respiration is also important for many microorganisms. Some, such as yeast, release alcohol as a by-product. Other microorganisms that respire anaerobically (anaerobes) produce lactic acid as a by-product, and the bacteria that live in the intestines of ruminants, such as sheep, cattle, and deer, generate methane. So, if energy can be generated anaerobically, why do organisms bother with oxygen and the challenge of gas exchange at all? This is because aerobic respiration is a far more efficient process; it produces 19 times more energy for each molecule of glucose.

How is oxygen produced?

Like animals, fungi, and most other organisms, plants and algae also liberate energy through aerobic respiration, in which they take in oxygen and release carbon dioxide

▼ GAS EXCHANGE IN DIFFERENT ORGANISMS

Gas exchange occurs in a variety of ways: diffusion directly across the cell surface or skin; across structures called papulae in animals such as starfish; along the tracheal system of insects; or across gills in fish or alveoli in the lungs of mammals.

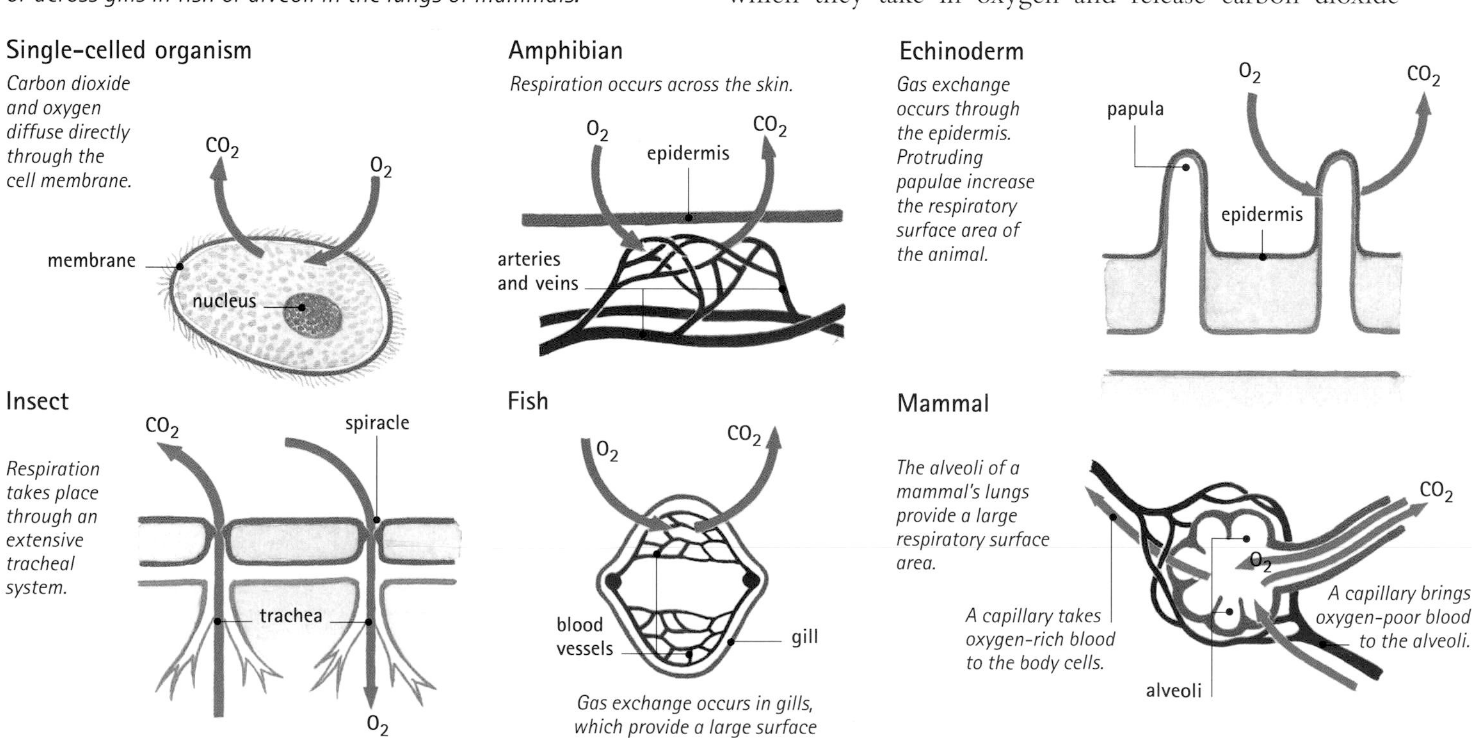

▲ *The elephant seal holds the record as the deepest-diving marine mammal, reaching depths of 5,180 feet (1,580 m) and routinely holding its breath for over an hour without ill effects.*

through pores called stomata. However, the movement of gases in these organisms changes over the course of a day. During daylight hours, plants take in carbon dioxide and release oxygen. They fix carbon dioxide and water to form sugars. This process, called photosynthesis, is driven by the energy of sunlight. The sugars are later stored or respired.

FEATURED SYSTEMS

BREATHING UNDERWATER Most water-living, or aquatic, animals breathe using gills—thin-walled respiratory surfaces across which gases diffuse into and from the water. *See pages 1016–1018.*

BREATHING AIR USING LUNGS Most four-legged animals depend on lungs for gas exchange. Lungs contain moist, thin-walled membranes and have a vast surface area across which gases diffuse. *See pages 1019–1020.*

THE TRACHEAL SYSTEM Insects, centipedes and millipedes, and many spiders use tracheae, tubes that take air through pores in the exoskeleton to the tissues where it is needed. *See pages 1021–1022.*

RESPIRATORY PIGMENTS These molecules occur in the blood of many animals and help carry gases around the animal's body. *See pages 1023–1025.*

VENTILATION Air or water in an animal's respiratory system can quickly become depleted of oxygen, so a continual flow is needed to bring in fresh supplies. *See pages 1026–1028.*

MAXIMIZING EFFICIENCY Animals and plants improve the efficiency of gas exchange by maximizing uptake rates, increasing surface areas, and avoiding water loss. *See pages 1029–1031.*

Oxygen is a by-product of photosynthesis and is released. Without photosynthesis, atmospheric oxygen would not occur and multicellular life as we know it could not exist.

What are the requirements for gas exchange?

Gas exchange depends on a process called diffusion. In this process, molecules move along a diffusion gradient—from a point of high concentration to a point of low concentration. Thus, for oxygen to diffuse across a respiratory surface, its concentration must be lower on the inside of a respiratory surface than on the outside.

Efficient gas exchange has some other requirements. The surface across which exchange takes place must be thin, to minimize the distance gases need to diffuse. It must also be moist: gases dissolve into the fluid. A moist respiratory surface is not a problem for aquatic organisms, because they are surrounded by water; but the need for moisture and the accompanying problems of water loss are important for terrestrial organisms. The final requirement is a large surface area; this requirement more than any other has shaped the evolution of respiratory structures.

Breathing through the skin

Microorganisms exchange gases through diffusion across their outer membrane. As oxygen is used up, more diffuses in; when carbon dioxide levels rise, some carbon dioxide diffuses out. Such diffusion across the outer membrane—

also called cutaneous gas exchange—is sufficient for some animals, especially those with most or all of their cells in contact with the outside, such as jellyfish and sponges. Many insect larvae (immature forms) breathe solely through their surface, as do many aquatic worms and sea slugs.

Changing shapes

For most larger animals, though, gas exchange across the skin cannot provide the body with enough oxygen. This is because the larger an animal is, the smaller its ratio of surface area to volume. The ratio is high for small animals; in other words, there is a lot of skin per unit of volume. However, for larger animals, the ratio gets progressively

COMPARATIVE ANATOMY

Giant skin breathers

Although cutaneous (across skin) gas exchange is generally ineffective for larger animals such as vertebrates, there are a few exceptions. Hibernating frogs draw oxygen through their skin as they doze on the bottom of ponds. Frogs use very little oxygen during this time, and diffusion across the skin can provide more than enough. Some salamanders and other amphibians called caecilians rely wholly on cutaneous gas exchange; these unusual amphibians do not possess lungs. However, they can live only in waters that are extremely rich in oxygen. The caecilian *Atretochoana eiselti* is by far the largest lungless terrestrial vertebrate, measuring up to 31 inches (80 cm) long. Not much is known about this amphibian, but it is believed to live in the cool, fast-flowing streams of western Brazil.

▼ *A whale's two nostrils are located on the top or back of the head and are more commonly called blowholes (toothed whales such as killer whales have a single blowhole). As the whale surfaces, the blowholes open and air is exhaled explosively through them, forming a misty vapor called the blow. Fresh air is then inhaled, and the blowholes close again.*

lower. The amount of skin available for gas exchange per unit volume decreases until a critical size is reached beyond which skin diffusion is too inefficient.

Larger animals need more efficient methods of oxygen uptake in which there is a greater surface area available for respiration. For example, starfish have projections over their surface called papules to increase surface area. Most aquatic animals, though, exchange gases through respiratory structures called gills. Gills can be internal, such as those of fish or mollusks. Other animals, such as young amphibians and mayflies, have external gills.

Gas exchange on land

Animals with gills rely on oxygen dissolved in the water. This reliance can be problematic: many aquatic insects, for example, are restricted to cool, fast-flowing streams that are high in oxygen. Land animals have access to far richer oxygen resources. However, respiratory surfaces need to be moist, so those of land animals are internal and can generally be shut off from the outside to minimize water loss.

EVOLUTION

The oxygen story

Oxygen is an extremely reactive gas. Oxides, chemicals formed by the reaction of oxygen with other elements, are all around us. Carbon dioxide, rust (iron oxide), and water (dihydrogen oxide) are examples. That there is any free oxygen in the atmosphere at all is due solely to the output of photosynthesis. Billions of years ago, the atmosphere contained little or no oxygen. The first organisms on Earth were anaerobic—they did not use oxygen to generate energy. Then, around 3.5 billion years ago, a group of bacteria called cyanobacteria appeared. These were the first photosynthesizers, and they had a dramatic impact. Atmospheric oxygen levels rose sharply. The oxygen caused an ecological disaster—almost all of the anaerobes were killed off. They survive today only in extreme habitats such as pond sludge or on hydrothermal vents. The composition of the air we breathe today is a legacy of this ancient apocalypse. Oxygen levels continued to rise, and life evolved to cope with the dramatic change in atmospheric composition. With oxygen (in the form of ozone) forming a layer in the upper atmosphere that cut out harmful ultraviolet (UV) light, more complex organisms began to evolve.

There are two main breathing strategies adopted by land animals. Almost all four-legged animals breathe through lungs. These are organs where air is alternately drawn in and expelled. Oxygen diffuses into blood carried by a network of capillaries. There, it binds to hemoglobin, an iron-containing protein in red blood cells. Hemoglobin also helps remove carbon dioxide, which is released at the lungs. Hemoglobin is a type of respiratory pigment. There are several others in the animal kingdom. Mollusks and crustaceans, for example, have a copper-containing pigment called hemocyanin that gives their blood a blue tint.

An alternative strategy

Insects have a completely different system. Air enters the body through a system of pores, called spiracles, in the exoskeleton. It passes through a series of increasingly small tubes called tracheae. The tips of the finest tubes are blind-ended and moist; oxygen diffuses from there into the tissues that need it. Therefore, almost all insects have no need for respiratory pigments. Myriapods such as centipedes also have tracheae. So do some spiders, although most also have capillary-rich folded structures called book lungs, with respiratory pigments to carry away the oxygen.

Breathing underwater

Aquatic animals need to take in dissolved oxygen that is in the water around them. Cutaneous gas exchange can provide enough oxygen for most tiny organisms. However, larger organisms must increase their surface area relative to their volume for this to prove efficient. Cutaneous gas exchange sometimes provides enough oxygen for larger animals with a very low metabolic rate (the rate of energy use by the body), such as hibernating frogs. Turtles, too, use cutaneous respiration to remain underwater for long periods. They pump water into the throat and rectum to take in more oxygen. However, most aquatic animals use respiratory structures called gills.

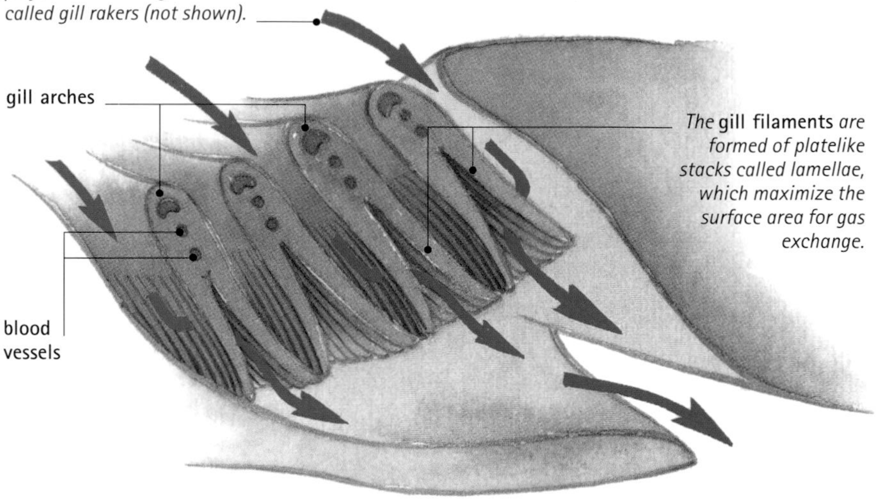

What are gills?

Gills consist of a stack of fine branches or folded flaps with a very thin outer membrane. The vastly increased surface area that gills provide allows rapid gas exchange to occur between the water and the inside of the gill's membrane. In most aquatic organisms, blood passes through the gills to take the oxygen away, often with the help of respiratory pigments.

Gills occur in various animal groups, from crustaceans and mollusks to vertebrates such as amphibians and fish. Some animals have internal gills. These can be folded structures such as fish gills, banks of filaments as in bivalves such as the giant clam, or branching trees as in sea cucumbers. Gills may also be external, as in polychaete worms and most aquatic insect

▲ GILLS
Trout
Each gill is made up of numerous gill filaments. The surface of each gill filament is folded into platelike structures called lamellae, creating a large surface area for gas exchange. Each lamella is supplied with blood vessels, which pick up oxygen from the moving water and release carbon dioxide.

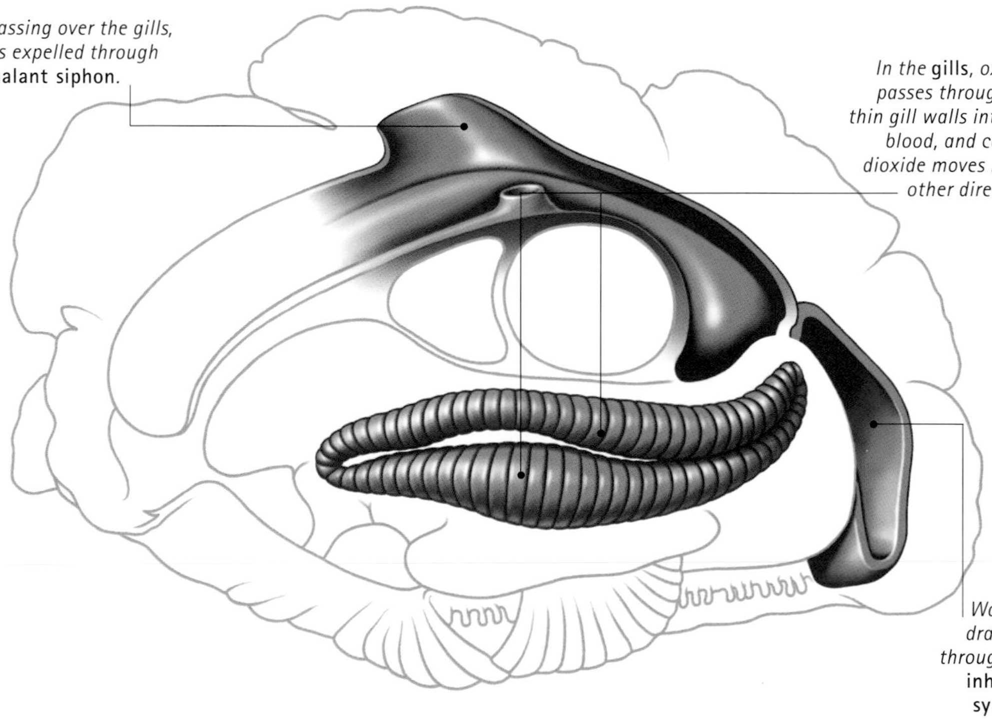

◀ GILLS
Giant clam
The giant clam draws water into its inhalant siphon and over its gills. There, oxygen passes from the water through the thin walls of the gills and into a series of thin tubes that carry blood around the clam's body.

COMPARATIVE ANATOMY

Breathing atmospheric air

Most aquatic four-legged animals must come to the surface frequently to breathe. The need to visit the surface may be a hindrance for these animals, but for many smaller organisms the surface provides a vital resource. It allows some to live in foul, low-oxygen waters where gills would be hopelessly ineffective. Rat-tailed maggots live in the sludge at the bottom of puddles. They breathe through a long, telescopic siphon that extends from their rear end. Some aquatic fly and beetle larvae in low-oxygen environments breathe by connecting their spiracles to airways inside the stems of submerged plants. Water scorpions live in cleaner waters, but they also breathe through a siphon. The siphon allows the insect to remain still as it waits to ambush prey.

▼ Rat-tailed maggot

The larva has an extensible tube, or siphon, through which it draws air into its body. The siphon enables the maggot to live in water that has a very low concentration of oxygen. Rat-tailed maggots feed on decaying plant material.

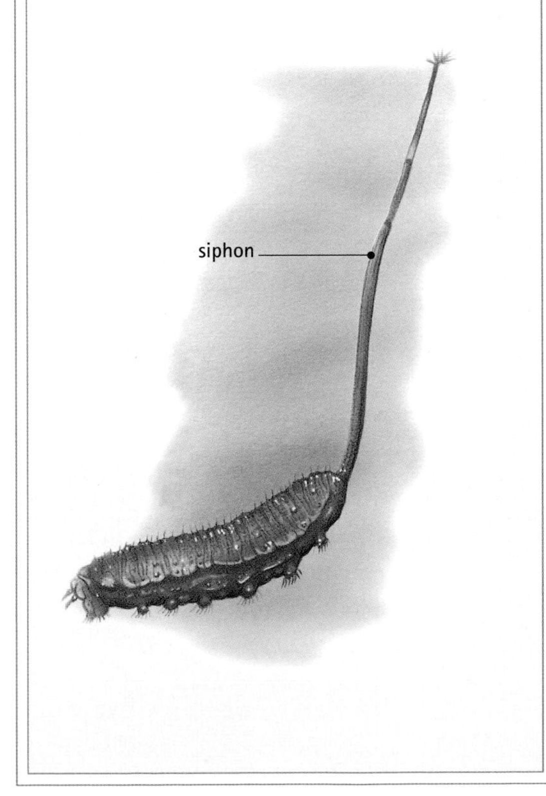

larvae. Insect gills connect directly to the tracheal system of air tubes. The gills of stonefly nymphs run along the body; those of young mayflies form tufts on the abdomen; and those of damselflies extend from the rectum.

Opposing streams

Water can be saturated with dissolved oxygen, but even then there is much less oxygen available than in the air above the surface. Oxygen levels drop further as the water temperature increases. So gills need to be superefficient to remove as much oxygen from the water as possible. Fish and many other gilled animals maximize uptake by using a countercurrent system. Water flows in one direction over the gills as the animal swims, for example, or by the production of a ventilating current. The blood inside the gill flows in the opposite direction. This ensures that blood takes in oxygen throughout its passage through the gills.

Other ways of getting oxygen

Gills are very common in the animal world, but they do not provide the only solution to getting oxygen underwater. Adult insects do not have gills. Many underwater insects must breathe air, but some store air so that they can minimize the time spent in the hazardous surface waters. These insects have hydrophobic (water-repelling) hairs that trap bubbles of air and hold them against the body. Backswimmers trap air bubbles using long hairs on their

COMPARATIVE ANATOMY

Gill form and function

Although gills evolved to facilitate efficient gas exchange, some animals have additional uses for their gills. Many use gills to help them swim. Young horseshoe crabs use their gills as paddles. Dragonfly nymphs have gills inside their rectum. These gills are ventilated by drawing in water; by squeezing the water out rapidly, the nymph can jet-propel itself forward. Gills may also be used for feeding. Bivalves such as mussels and clams use their gills (or ctenidia) to filter small particles from the water. The particles are sorted by appendages near the mouth called labial palps before being passed into the mouth. Some gills are not used for gas exchange at all. Mayfly nymphs use theirs to beat a current of water over a patch of folded cuticle on their abdomen; gas exchange takes place there instead.

IN FOCUS

Underwater arachnids

Spiders are an amazingly diverse group of invertebrates. However, very few spiders have evolved for a life in water. A few hunt on the surface film of ponds and pools, but just one species spends its life underwater—the water spider. Lacking gills or paddles, it may seem ill-suited for an aquatic life, but it has an amazing method for getting the oxygen it needs to respire. The spider weaves a sheet of silk and attaches it to underwater plants; this forms a domed "diving bell." The spider's abdomen is covered by hairs that trap air when the animal visits the surface. The spider uses its legs to flick air bubbles into the diving bell.

The water spider rests in its diving bell during the day. At night it sallies out to catch aquatic insects and small fish, beating its legs to drive itself through the water. A thin layer of air around its body acts as a physical gill. Prey is taken back to the diving bell and eaten.

▶ SPINNING A WEB *A water spider spins an underwater sheetlike web between the fronds of an aquatic plant. The spider then surfaces to collect air on its hairy abdomen.*

▶ FILLING WITH AIR *The spider uses its legs to flick off the air on its abdomen into the web to form a nest with a bell-shaped bubble. The spider rests in the nest during the day, but hunts at night, eating prey such as tadpoles always in the nest.*

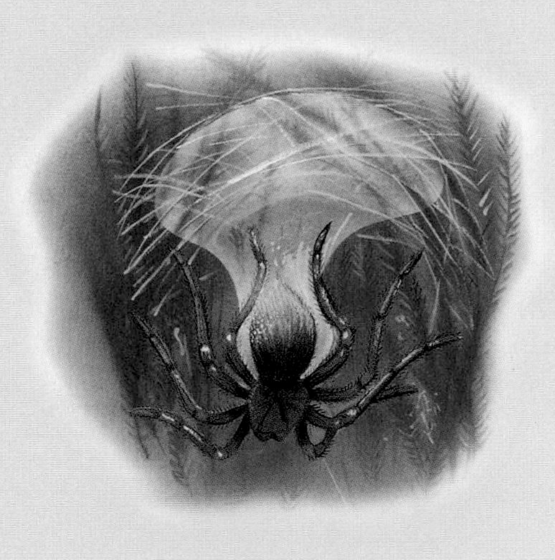

underside, as well as under the wings and on the upper surface of the forewings, where their spiracles (airholes) are located. Bubbles can last a long time before they run out of oxygen. This is because as the insect uses up oxygen, more diffuses in from the surrounding water. Such air bubbles are called physical gills.

Some insects, such as *Aphelocheirus* bugs, can remain submerged indefinitely. These bugs have a very dense layer of short, bent hydrophobic hairs coating much of the body. The hairs hold an extremely thin layer of air close to the body. This is called a plastron. It acts as a gill, drawing oxygen from the surrounding water.

Breathing air using lungs

Gills do not work effectively on land. The gill filaments clump together, reducing the surface area available for gas exchange. All land vertebrates, instead, breathe through organs called lungs. These organs contain chambers that provide a thin, moist surface of vast area ideal for gas exchange. Amphibians such as frogs have large but short lungs with a single chamber, although the walls are partitioned. In reptiles, partitioning is more elaborate, and some reptiles, such as snakes, get by with just one lung. The most complex lungs occur in the vertebrates that have the highest demand for oxygen, namely the mammals and birds.

COMPARATIVE ANATOMY

How snails breathe

Lungs do not occur solely in vertebrates—some invertebrates have them, too. Land snails evolved from gilled marine ancestors, but they have lost their gills entirely. Instead, their mantle cavity—a space in the snail—has evolved into a lung. Its walls are richly supplied with blood vessels and are strongly ridged to increase the surface area. Some land snails have secondarily become aquatic, living in freshwater. Most must visit the surface regularly to fill their lung, although species that live in oxygen-rich waters, such as pond snails, fill the lung with water and rely on the diffusion of dissolved oxygen across its walls.

The passage of air

Air normally enters mammals through the nostrils. This allows breathing and feeding to occur simultaneously, although the linkage with the mouth usually remains. Breathing through the nostrils also allows the nasal hairs to filter the air and structures called nasal turbinates to warm and moisten it.

A flap of tissue called the glottis tilts up to allow air into the larynx, which contains the vocal cords. In swallowing, the glottis tilts down, so food passes into the esophagus

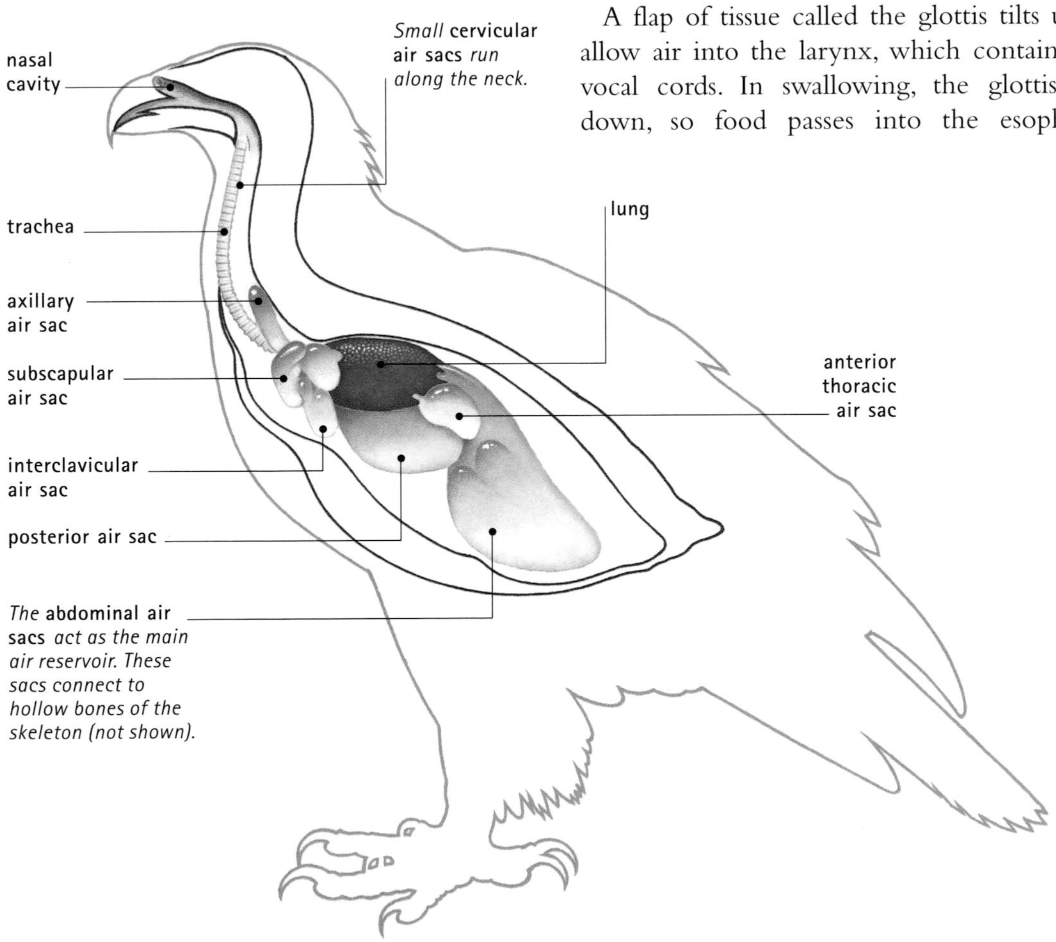

◀ LUNGS AND AIR SACS
Bald eagle
Birds have lungs and also a complex series of air sacs. The lungs and air sacs, along with hollow bones linked to air sacs, allow a continuous stream of air to pass through the lungs in a one-way flow.

EVOLUTION

Fish with lungs

Tetrapods, or four-legged vertebrates, descended from lobe-finned fish that emerged from the sea about 350 million years ago. These ancient fish must have been able to breathe atmospheric air. There are a few species of fish that can do this today. Some characins can use their swim bladders as a lung, and *Hoplosternum* catfish use part of their gut. The best-known air-breathing fish are the lungfish. Their lungs are derived from the swim bladder and may be similar to those of the earliest tetrapods. Lungfish lungs are divided into pockets richly supplied by blood vessels. Some lungfish have just a single lung; it helps them get enough oxygen in times of shortage so they can survive in oxygen-poor waters. Others have two lungs, and these fish hole up in mud cocoons when their lakes dry up, breathing occasionally through their lungs until the rains return.

▼ *Lungfish lungs evolved from the swim bladder. The fish use their lungs for breathing in seasonal droughts. There are three types of lungfish: South American, African, and Australian. The African and South American species have two lungs, and the Australian (below) has just one lung.*

instead. Air passes through the larynx into the trachea. The trachea is made of smooth muscle reinforced by rings of cartilage. Mucus-producing cells line the trachea; the mucus traps particles, and banks of tiny filaments called cilia carry the mucus up to the throat. At its lower end, the trachea divides into two main airways, each supplying a different lung. These are called the bronchi.

The lungs are in the chest cavity, where they are protected by the rib cage. The bronchus of each lung divides into thousands of smaller tubes called bronchioles. These lead to tiny air sacs, or alveoli. Richly supplied by capillaries, the thin-walled alveoli are the site of gas exchange. The human lung contains millions of alveoli; they give a total surface area for respiration of about 800 square feet (75 sq m).

How is breathing controlled?

The movement of air into and out of the lungs is involuntary, or automatic. The breathing control center is in the medulla, part of the brain stem. Neurons (nerve cells) send signals to the rib and diaphragm muscles to regulate the rate and depth of breathing.

Breathing is affected by a number of factors. Receptors in the aorta (the body's main artery) and carotid (neck) arteries monitor amounts of oxygen and carbon dioxide in the blood and also its acidity. The medulla itself measures levels of carbon dioxide in the fluid surrounding the brain and spinal cord. Receptors that respond to mechanical stimuli (mechanoreceptors) in the lungs monitor how stretched the tissues are, and this prevents the lungs from overinflating. Receptors in the lungs and airways transmit messages to the medulla if irritating particles or chemicals are present. This triggers a sudden exhalation in the form of a cough or sneeze.

Super lungs

Flight is the most energetically demanding form of locomotion. Birds' lungs are very different structurally from those of mammals, and they are also far more efficient. A bird's trachea divides into two bronchi as in mammals, but inside the lung each bronchus separates into two main airways linked by smaller channels called parabronchi. Gas exchange takes place through the walls of the parabronchi, which have a unique structure. They are permeated by fine, branching tubes called air capillaries, along which gases diffuse. The tubes are intertwined with blood vessels.

Connected to the lungs are several air sacs, which fill along with the lungs when the bird breathes in. These are also linked to hollow cavities inside some of the bones. Air sacs act to store fresh air; the air moves into the lungs when the bird breathes out. In this way, birds' lungs can absorb oxygen throughout their respiratory cycle.

The tracheal system

Lungs provide one solution to the challenge of exchanging gases on land. An alternative solution is to have a system of air tubes that carries oxygen directly to the tissues; this is called a tracheal system. Several invertebrate groups have a tracheal system, including the velvet worms, myriapods, most spiders, and the largest animal group of all, the insects. The insect tracheal system is supremely efficient; it needs to be, because insect flight muscle uses oxygen faster than any other known tissue.

How the tracheal system works

Paired pores called spiracles are the gatekeepers of the tracheal system; oxygen and carbon dioxide move in and out through them. Insects such as fleas, bees, and moth larvae have eight or more pairs running along the body; other insects, such as some fly larvae, have just one functional pair, or even, in the case of some aquatic species, none at all. Spiracles are usually kept closed to reduce water loss, and they often contain a hair-fringed plate to keep out dirt.

IN FOCUS

Breathing inside other animals

Many insects develop inside other animals, and some of these parasites go on to kill their hosts. How do these insects get the air they need? Chalcid wasp larvae, which are internal parasites, or endoparasites, of other arthropods, connect to the air outside through a tube that formed part of their egg. The tube penetrates the host's body wall. Some tachinid fly larvae force their arthropod hosts to grow a structure called a respiratory funnel. This is an envelope of tissue that surrounds the larva and plugs into the host's spiracles. Many endoparasites breathe atmospheric air directly. For example, human botflies burrow into flesh but keep an airway to the outside open through the skin.

Each spiracle opens into a chamber called an atrium, from which the tracheae arise. The tracheae are large, cuticle-lined tubes that carry gases between the spiracles and tissues. Spirals of thick cuticle called taenidia strengthen the tracheae and keep them from collapsing if the

◀ *The tracheal system of insects and other land-living arthropods becomes less efficient with increasing size, and therefore limits the size of the animal. The goliath beetle from Africa is the heaviest known insect. It weighs up to about 3 oz (85 g) —which is more than some birds. However, in prehistoric times, some insects, such as dragonflies, were much larger because there was more oxygen in the air then.*

EVOLUTION

The constraints of tracheae

For small organisms, tracheae are vastly superior to lungs. Tracheae deliver oxygen efficiently, and energy is not squandered on powerful pumping hearts or respiratory pigments. However, tracheal systems become less efficient with increasing size, placing a constraint on the maximum size of insects; that is why there are no rhino-size beetles or bird-hunting dragonflies.

Reliance on the tracheal system is also responsible for insects' inability to conquer marine habitats. Just one group lives at sea, the sea skaters, and its members only glide on the surface. The incompressibility of the tracheal system renders aquatic insects buoyant; a marine insect would not be able to dive quickly enough to escape a predator.

pressure drops. In certain parts of the tracheae, the taenidia are absent, and the tracheae expand to form collapsible air sacs. These sacs allow ventilation of the system.

Tracheae branch into increasingly narrow airways. Eventually they become fine, blind-ended tubes called tracheoles. Every body cell is close to or in contact with a tracheole. Oxygen diffuses through the thin tracheole membrane and dissolves into the fluid surrounding the cells. Oxygen then enters the cells, with carbon dioxide moving in the opposite direction.

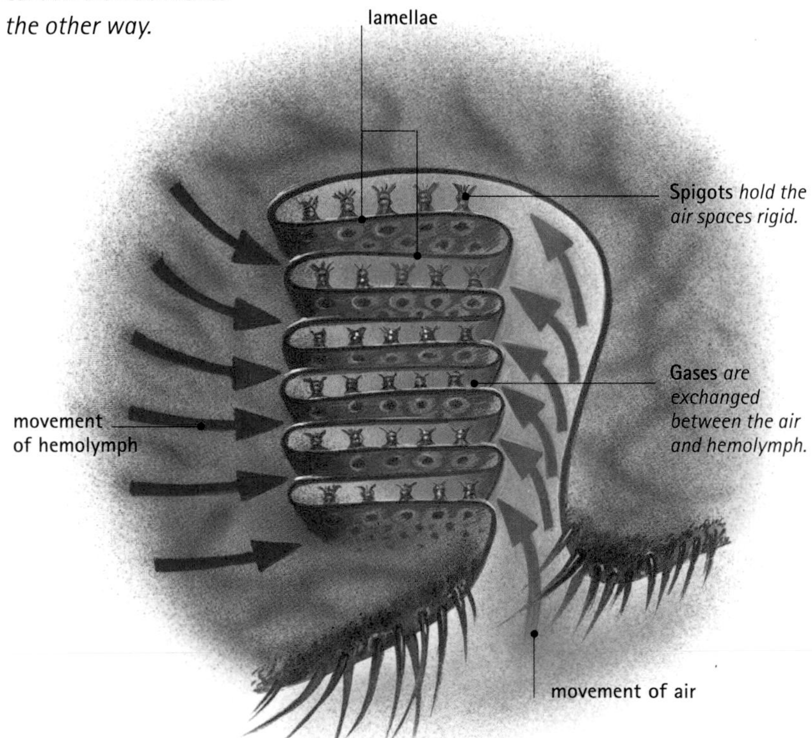

▼ **BOOK LUNGS**
Most spiders and scorpions have respiratory organs called book lungs. Oxygen from the air moves across the platelike lamellae into the bloodlike hemolymph; waste carbon dioxide moves the other way.

Where are these tubes?

Tracheae from neighboring spiracles link to form a pair of major trunks that run the length of each side of the body. Smaller tubes, called tracheoles, branch off to supply the organs and tissues, and they also link the main trunks together. Tracheoles are most abundant in regions of high oxygen demand, such as the wing muscles. During flight, the thorax tracheae are cut off from the rest of the respiratory system. This isolation stops the oxygen requirements of the flight muscles from depriving other essential tissues, such as the brain.

Underwater tracheae

Aquatic insects use tracheae, too. The spiracles of insects that breach the surface to get air, such as backswimmers, are surrounded by hairs that repel water strongly. The spiracles of mosquito larvae bear valves that close when the animal dives. Often only the spiracles near the end of the body are functional in young aquatic insects that visit the surface, although more become active as the insect approaches its final molt. The gills of young caddis flies and damselflies are richly supplied by tracheoles, while the spiracles of some water beetles and bugs open directly into bubbles or films of air on the body.

Spider breathing

Most spiders also have tracheae, although they are less efficient than insect tracheae and work in a different way. They do not deliver oxygen directly to cells; instead, they release oxygen into the bloodlike hemolymph, where it binds to a respiratory pigment. Spider tracheae evolved from structures called book lungs. Book lungs occur in the abdomen. A small atrium enlarges to form a series of horizontal air pockets that are interwoven between stacks of thin-walled leaflets (lamellae) through which hemolymph flows.

Mygalomorphs (tarantulas and relatives) have two pairs of book lungs; in most other spiders, the hind pair has evolved into tracheae, whereas a few groups of spiders have lost their book lungs altogether. The tracheae allow large volumes of air to be stored, and also limit the need for a powerful circulatory system.

Respiratory pigments

In most animals, oxygen binds at the respiratory surface to a molecule called a respiratory pigment. Respiratory pigments link readily to oxygen where it is freely available: in the lungs, for example. This binding property is called oxygen affinity. The binding is reversible, allowing oxygen to be released wherever it is needed for respiration.

Amazing hemoglobin

Hemoglobin is the most widespread respiratory pigment in the animal kingdom, occurring in almost all vertebrates and in invertebrates such as earthworms and echinoderms. It is also by far the most efficient. A hemoglobin molecule consists of four long protein chains, or globins, each coiled into a kidney shape. In the middle of each globin is a molecule with an ion (minute particle) of iron at its center. This molecule is called a heme group. The arrangement of globins permits four oxygen molecules to be carried at any one time. Hemoglobin allows 50 times more oxygen to be transported around the body than could be carried by the plasma (liquid part of the blood) alone.

Path of hemoglobin

The oxygen affinity of hemoglobin is highest in the lungs. There, it is cool and plenty of free oxygen is readily available. The hemoglobin binds tightly to the oxygen and changes from purple red to bright red in the process. How does hemoglobin release oxygen to the tissues that need it? Much depends on the pH, or relative acidity, which is determined by the amount of tiny hydrogen ions present.

As a cell respires, it produces carbon dioxide. An enzyme called carbonic anhydrase converts much of the carbon dioxide into carbonic acid. This acid dissolves into the plasma, forming

CONNECTIONS

COMPARE the respiratory pigment hemoglobin in verterbrates, such as the ***NEWT***, ***PUMA***, and ***TORTOISE***, with hemoglobin in the invertebrate ***EARTHWORM***. In vertebrates, hemoglobin occurs inside red blood cells, whereas in the earthworm, the hemoglobin is free in the blood.

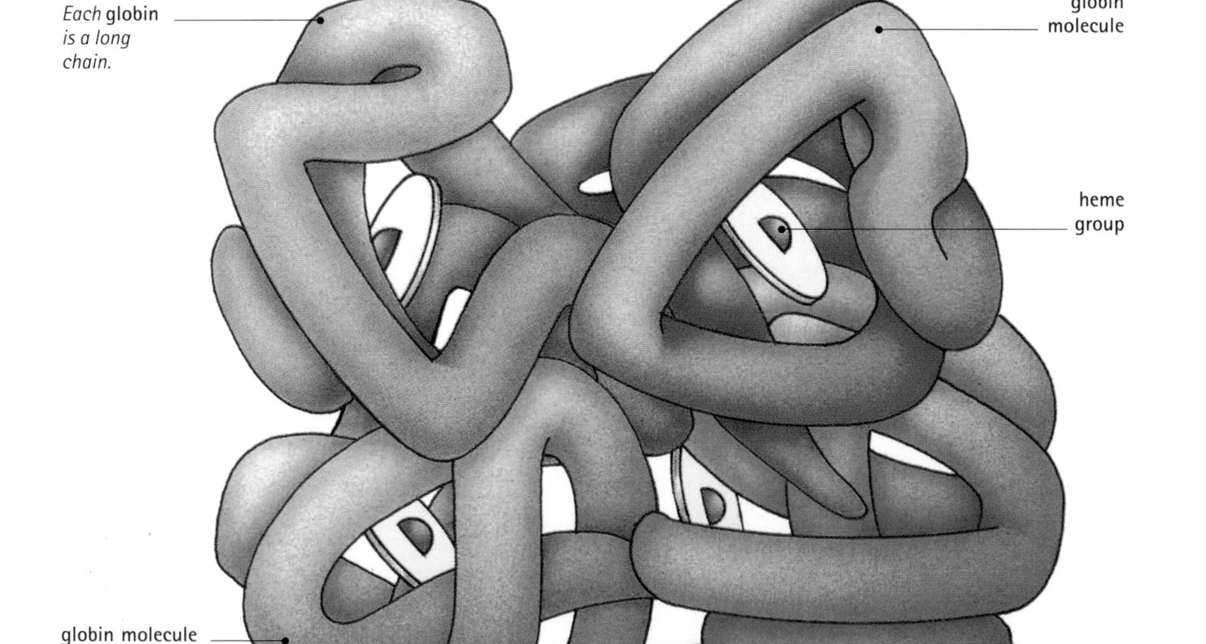

◀ HEMOGLOBIN
Hemoglobin is the respiratory pigment of vertebrates and some invertebrates, including earthworms and starfish. Each hemoglobin molecule is made up of four globin molecules. Each globin has an iron-containing heme group, which transports oxygen and gives blood its red color.

IN FOCUS

Other vertebrate pigments

Hemoglobin is not the only vertebrate respiratory pigment. A similar pigment, called myoglobin, occurs in muscles. Myoglobin serves as an oxygen store. It binds extremely tightly to oxygen, releasing its cargo when oxygen shortages occur during exercise. Babies in the uterus have a different type of hemoglobin from their mothers. Called fetal hemoglobin, it has a very high oxygen affinity; this allows it to take oxygen from the mother's blood at the placenta.

bicarbonate and hydrogen ions. The hydrogen ions increase the acidity. An arriving hemoglobin molecule meets a warm, oxygen-poor, and—crucially—acidic environment. This combination of factors causes the molecule's oxygen affinity to decrease. The hemoglobin's essential cargo of oxygen is released, diffusing into the cell, and respiration takes place. The influence of the acidity caused by carbon dioxide on hemoglobin is called the Bohr effect.

Helping remove carbon dioxide

The principal role of hemoglobin is oxygen transport, but it plays an important part in the removal of carbon dioxide, too. There are several ways of getting carbon dioxide to the lungs. A small amount of the gas dissolves directly into

▼ *Red blood cells, or erythrocytes, contain the respiratory pigment hemoglobin. In humans and other mammals, the cells are disk-shape and lack a cell nucleus. In other vertebrates, the cells are oval and have a nucleus (x 2,000).*

COMPARATIVE ANATOMY

Life without pigments

Not all animals need respiratory pigments. Insects have tracheae that deliver oxygen directly to cells, so they do not need pigments. However, there are a few exceptions. Backswimmers are aquatic bugs that use hemoglobin as an oxygen store, releasing the gas to help them maintain buoyancy in the water. Just one group of vertebrates does without hemoglobin: the icefish. Icefish rely on dissolved oxygen in the plasma. To get enough oxygen to the tissues, icefish have a massive heart that keeps large amounts of blood at a very high pressure.

the blood plasma. About half of the rest binds to hemoglobin, and the remainder—converted into carbonic acid in red blood cells—is in the form of ions: hydrogen ions and bicarbonate ions. Hydrogen ions bind to the protein part of hemoglobin; bicarbonate ions are transported out of the red blood cell into the blood plasma. Both sets of ions are carried to the lungs; there, carbonic anhydrase re-forms them into carbon dioxide, which diffuses away.

A riot of colors

Vertebrate hemoglobin is contained within red blood cells, but in many other groups, such as annelid worms, hemoglobin drifts freely in the blood. In the animal kingdom there are many other respiratory pigments besides hemoglobin. Tunicates, also called sea squirts, are invertebrate chordates and are close relatives of vertebrates. However, they use a respiratory pigment that differs radically from hemoglobin. The pigment contains the metal vanadium. Contained within blood cells called vanadocytes, the pigment gives the blood a bright green color.

The copper-containing pigment hemocyanin occurs in the bloodlike hemolymph of many mollusks and in arthropods such as crustaceans and arachnids. Hemocyanin-containing blood is blue when it is carrying oxygen and colorless otherwise. Hemocyanin is always free in the blood and not contained within cells (unlike vertebrate hemoglobin, which is contained in red blood cells). However, hemocyanin is only about one-quarter as efficient a carrier of oxygen as hemoglobin.

Many respiratory pigments, like hemoglobin, contain ions of the metal iron. However, the colors of the pigments can be very different. For example, chlorocruorin is a large molecule that is found in the blood of tube-building polychaete worms, such as *Sabella* fan worms. Like hemoglobin, chlorocruorin is deep red when oxygen-laden, but it becomes green when it is deoxygenated.

▲ *Some fan worms have an iron-containing respiratory pigment called chlorocruorin. This pigment, like hemoglobin, is red when carrying oxygen, but it turns green when it gives up the oxygen.*

Ventilation

CONNECTIONS

COMPARE ventilation in land mammals, such as the ***RHINOCEROS***, with that in bony fish such as the ***SAILFISH***. Mammals use their ribs and diaphragm to draw air into and out of the lungs, whereas in bony fish water enters the mouth and is pumped or pushed across the gills, which pick up oxygen from the water.

Without a diffusion gradient, gas exchange cannot take place. So animals need to circulate fresh water or air continually over the respiratory surfaces and move out the air that is poor in oxygen and rich in carbon dioxide. This circulation process is called ventilation.

For some organisms, shape alone provides the necessary ventilation without the need for any extra energy expenditure. Mountain midges are flies that pupate (become adult) in streams. The pupae have a shape that creates vortices, or swirls, in the current. The vortices cause a decrease in pressure that makes oxygen collect as bubbles on the insects' gills.

Sponges are riddled with pores; they join to form a central chamber that empties at the osculum, a large hole at the top of the animal. The shape of a sponge allows it to exploit a physical property of fluids to ventilate itself for free. Fast-moving water creates a lower pressure than slow-moving water. This phenomenon is called the Bernoulli effect. Water moves more slowly at the ocean floor, owing to friction. The water even a few inches above the ocean floor, where the osculum empties, moves slightly quicker. The lower pressure draws water up through the sponge's pores to the osculum. Sponges generally do not live in still waters; they need some water movement to power this respiratory current.

Some land animals also rely on the Bernoulli effect to ventilate their burrows. Prairie dogs, for example, build vast networks of tunnels with openings at different altitudes. Air is sucked through the higher vents from the lower ones to ventilate the network.

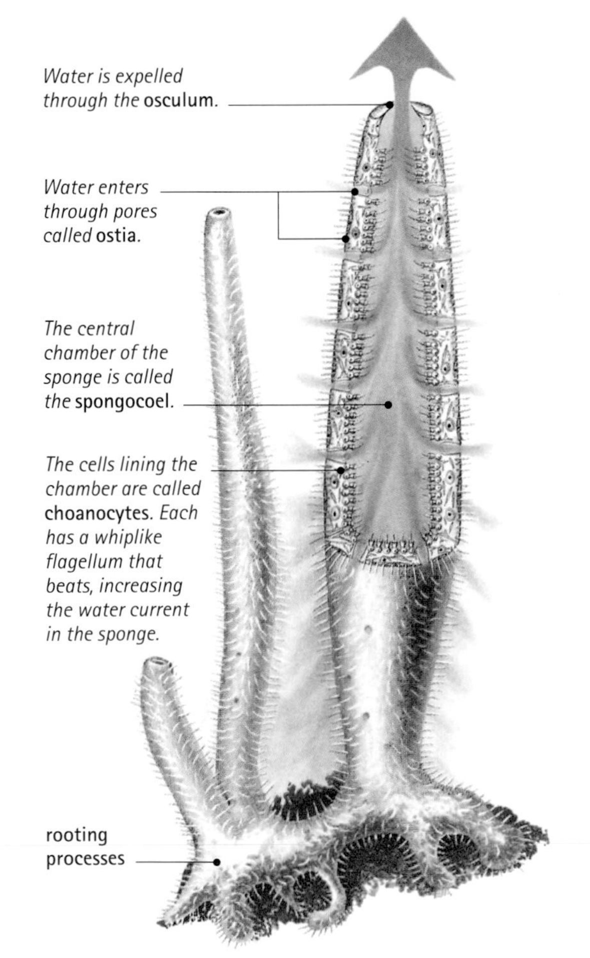

▶ VENTILATION IN SPONGES
Sponges are covered with pores called ostia, through which water enters a central chamber before exiting through the osculum at the top of the sponge. Pressure is lower in the faster-moving water around the osculum than at the base of the sponge, where the water moves slowly. The pressure difference creates a constant stream of water that ventilates the inside of the sponge.

Ventilating gills

Most animals must create their own ventilation currents. Crustaceans such as amphipods and many burrowing shrimp flap segments at the end of the abdomen called pleopods to ventilate their gills. Lugworms bring fresh water into their burrows by passing rhythmic waves along their body. Bivalves have banks of tiny filaments called cilia on their ctenidia (gills); they beat back and

COMPARATIVE ANATOMY

Ventilating trachea

Insects need to ventilate their tracheae, but these are kept rigid by taenidia and are largely incompressible. However, there are sections called air sacs that are not stiffened by cuticle. Compression of these air sacs flushes the system. To compress the air sacs, many insects reduce the volume of the abdomen. This can be done by telescoping the segments or by squeezing the upper surface downward. Either way, the pressure of the hemolymph inside rises, squashing the air sacs.

IN FOCUS

Ventilating gills at speed

Most sharks ventilate their gills in a fashion similar to teleosts, although water may enter through both the mouth and the spiracle, the modified first gill chamber. Water is then forced across the gills before passing out through the gill slits. These sharks can gather oxygen continuously. Some, such as the great white shark, and many fast-swimming teleosts such as mackerel and tuna, do not waste energy by pumping water through the gills. Instead, these fish rely on moving fast relative to the water. Fast movement forces water through the gills, a process that biologists call ram ventilation. Ram ventilators must swim constantly to exchange gases efficiently.

forth to create a current. Cephalopods such as octopuses use oxygen more rapidly than most other aquatic invertebrates and require stronger currents. Cephalopods rely on muscles to open and close the mantle cavity, which is divided into inhalant and exhalant chambers. The muscular activity draws water through a slit, over the octopus ctenidia, and out through a funnel.

Bony fish, or teleosts, such as goldfish also actively pump water over their gills, again, in one direction. Water is drawn into the mouth, which then closes to force it into the gill chambers. They force it across the gills and out. A flap of tissue called the operculum covers the gills, protecting them and closing them off when the mouth is open.

Ventilation in amphibians

Gills are ventilated in one direction, but lungs are ventilated tidally, with air moving in and out along the same path. Lung ventilation is powered in a variety of ways. Amphibians breathe through their nostrils. When the nostrils are open, a flap of tissue called the glottis covers the entrance to the larynx. The floor of the mouth pushes down to suck air into the mouth. The glottis then opens and the lung walls contract, forcing depleted air into the mouth. The nostrils close, and the floor of the mouth rises to force air back into the lungs. This system may not seem efficient, since air from the lungs mixes with fresh air coming in. However, between cycles the amphibian keeps the glottis closed and nostrils open, and moves the floor of the mouth up and down. This action removes the last of the depleted air from the last cycle. Also, since they have a moist skin, almost all of the carbon dioxide produced by amphibians can be removed through the skin. Exhaled air contains much less of this gas.

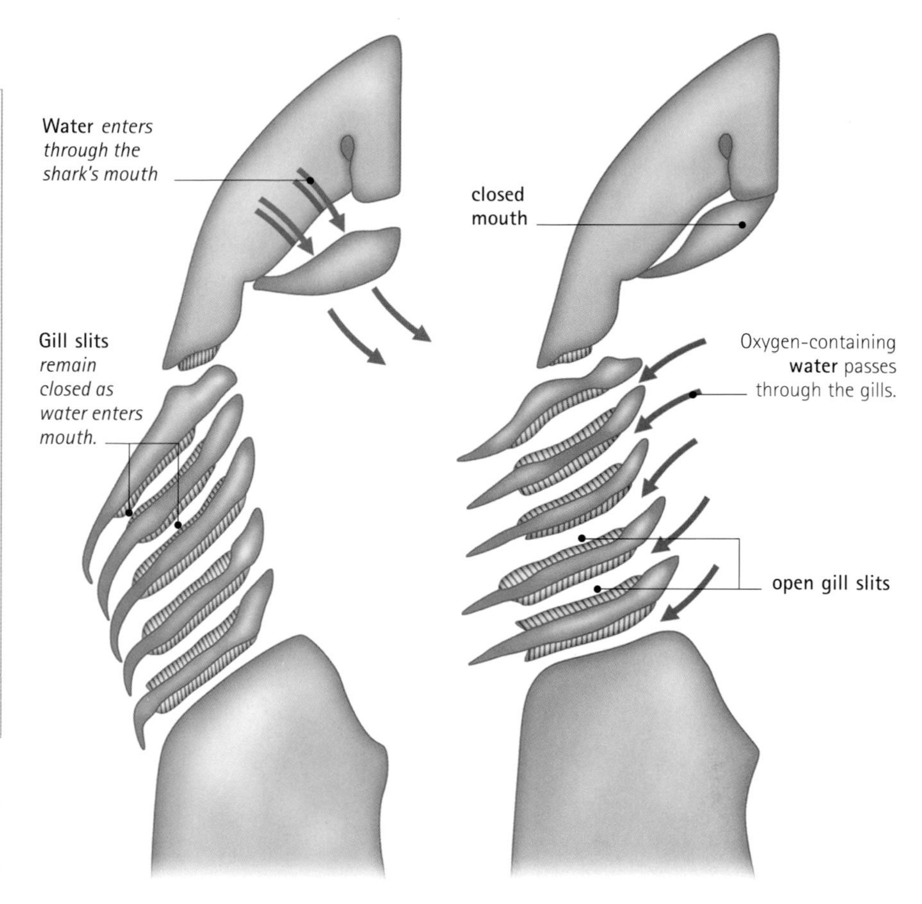

▲ VENTILATING GILLS

Shark (cross section of lower jaw)

Water enters the shark's open mouth (left) when the gills are closed. The shark then closes its mouth and opens the gill slits, allowing the oxygen water to pass across the gills and out through the gill slits (right).

Ventilation in reptiles

Reptiles cannot get rid of carbon dioxide through their waterproof skin. Therefore, they must change the air in their lungs frequently to remove as much carbon dioxide as possible. Gas exchange takes place in the front part of a reptile's lungs; the hind part acts as a bellows. Most reptiles breathe in by expanding the rib

COMPARATIVE ANATOMY

Ventilation in mammals and birds

Mammals fill and empty their lungs using a combination of muscles. To breathe in, the diaphragm and intercostal (rib) muscles contract, swinging the ribs outward and expanding the chest cavity. This action lowers the pressure inside the cavity, so air flows in to inflate the lungs. Breathing out occurs as these muscles relax; it is aided by the elasticity of the lung tissue, which contains molecules called elastins that help the tissue recoil. Abdominal muscles also contract to force more air out of the lungs during exercise.

Birds have a very different respiratory system from mammals. When a bird inhales by raising its sternum (breastbone) air fills the lungs and also a system of air sacs. As the bird exhales, oxygen-rich air from the air sacs is forced through the bird's lungs.

▼ VENTILATION IN HUMANS

On inhalation, the diaphragm and rib muscles contract, expanding the rib cage and drawing air into the lungs. The muscles then relax, contracting the rib cage, and air is exhaled.

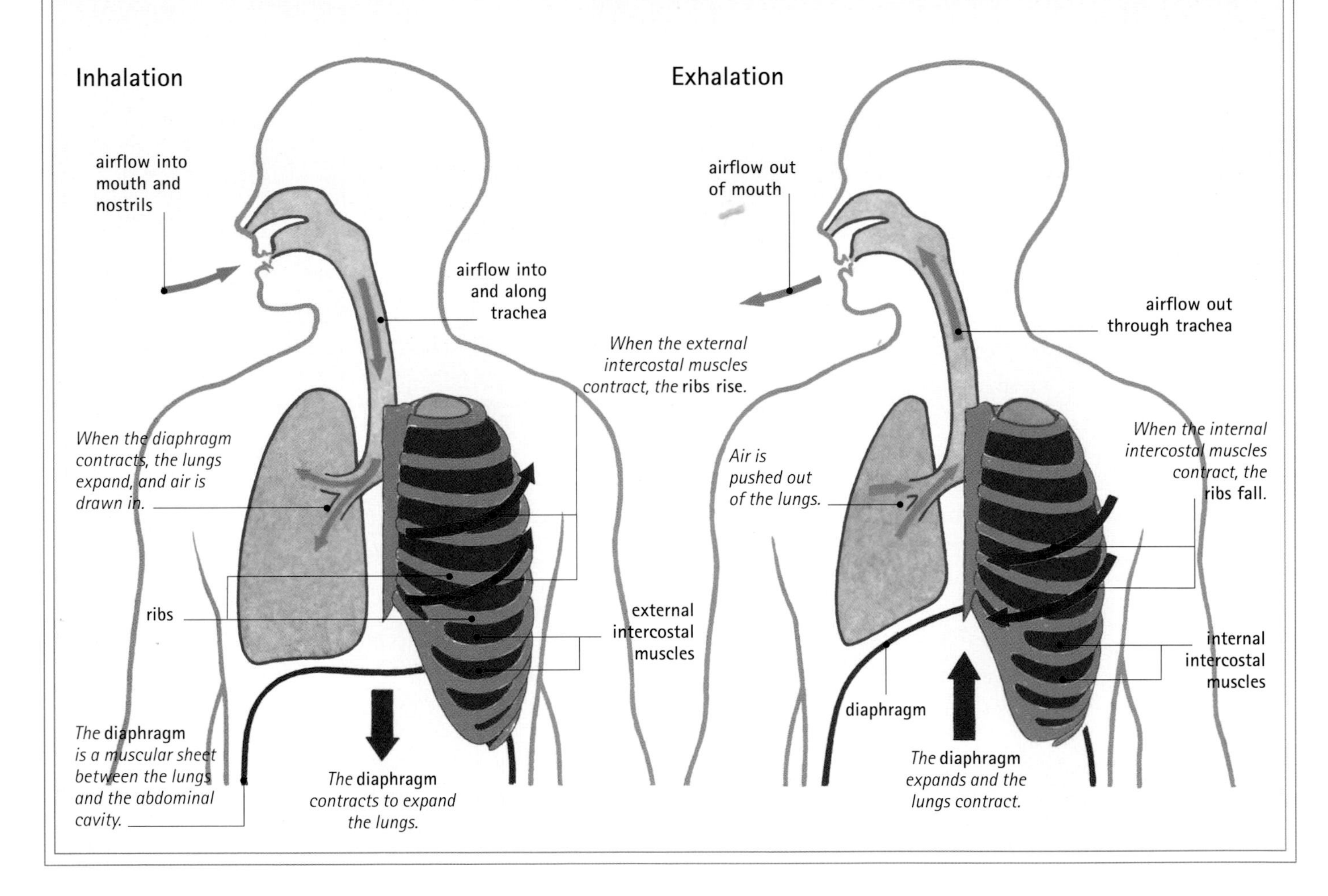

cage; this expands the lungs, causing a drop in pressure so air flows in. The lung tissues and ribs of reptiles are elastic, and this recoil powers the exhalation of the air.

However, ventilation is different in turtles. Their ribs are fixed in place to form part of the shell; turtles rely instead on a sheet of tissue called a diaphragm that connects to the limbs. Extension of the limbs pulls the diaphragm down, causing air to move into the lungs. Exhalation is then caused by limb retraction. Crocodiles use neither ribs nor a diaphragm but a muscle called the diaphragmaticus. This attaches the pelvis to the liver, which is linked to the lungs. When the diaphragmaticus contracts, the lungs expand, causing them to inflate. Breathing out is caused by the contraction of abdominal muscles that pull the liver forward.

Maximizing efficiency

Organsims have a variety of features that maximize their respiratory efficiency. Skin breathers can increase their surface area relative to volume by becoming long in one direction; in other words, by evolving into a wormlike shape. Folds, bumps, or depressions on the skin increase surface area, too. Gills and lungs follow this principle.

CLOSE-UP

Abandoning the nucleus

Mammalian red blood cells have a unique property that allows them to maximize the amount of oxygen in the blood. Mature red blood cells lack a nucleus, the control center containing genetic material that occurs in all other body cells. This allows more oxygen-carrying hemoglobin to be packed in.

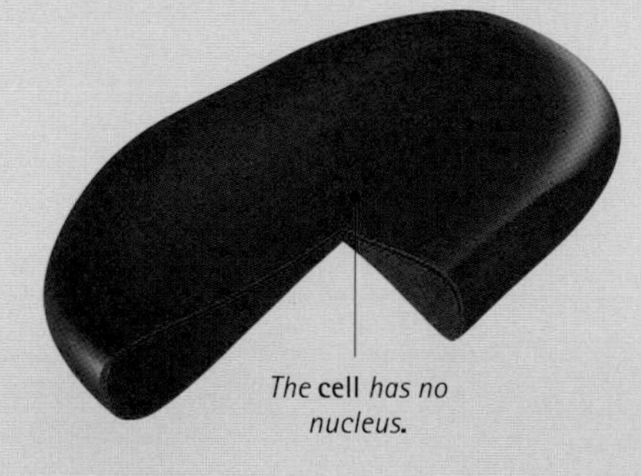

▲ Oxygenated mammalian red blood cell
Mammalian red blood cells do not have a nucleus. When laden with oxygen from the lungs, the hemoglobin-packed red blood cells are bright red.

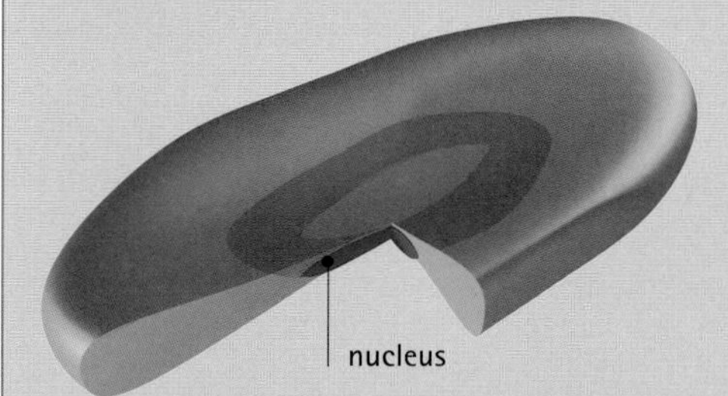

▲ Deoxygenated lizard red blood cell
Lizards have flatter, more oval, red blood cells with a nucleus. Without oxygen, the iron-containing hemoglobin is blue.

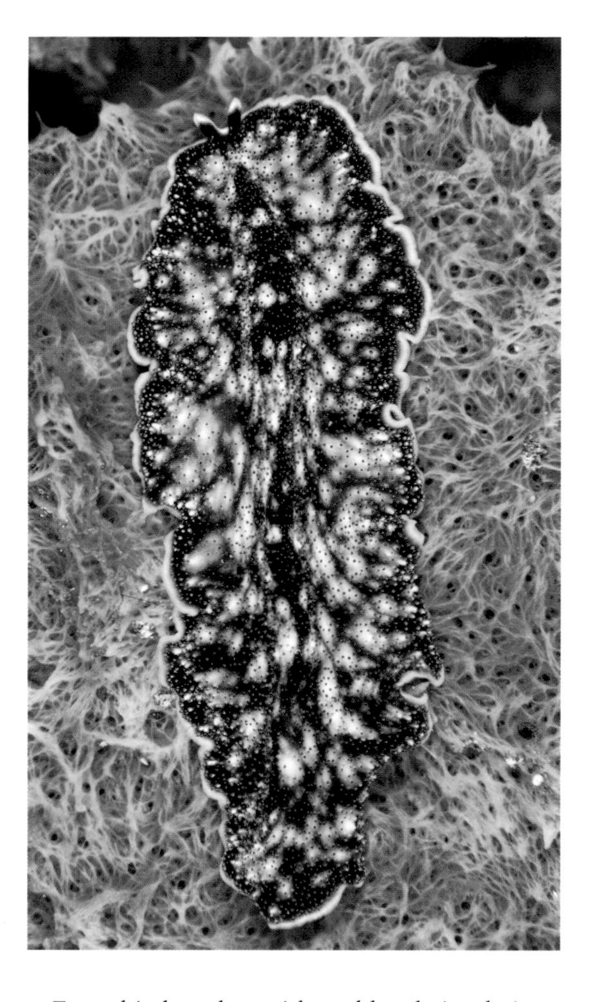

◀ *This marine flatworm does not have any specialized respiratory organs such as gills. It relies solely on the diffusion of oxygen from the seawater across the worm's large surface area. Because the worm is flat, the oxygen does not have to travel far to reach all respiring cells.*

For a skin breather without blood circulation, body parts cannot be situated far from the skin; otherwise, oxygen will not diffuse fast enough to reach them. For example, flatworms and the fronds of the seaweed kelp are long and flat to minimize this diffusion distance. Insects can alter the diffusion distance at the tips of their tracheoles; these tips contain fluid that is withdrawn during periods of activity and returned when the insect is at rest.

Improving the diffusion gradient

The most efficient way to maintain a diffusion gradient is through a countercurrent system, as in fish and cephalopods. Water and blood move in opposite directions; when blood first enters the gill, it is almost devoid of oxygen, but the water it encounters has a little more. As the blood progresses, it encounters richer water, but there is always less oxygen in the blood than in the water. A diffusion gradient is thus permanently maintained. Teleost gills remove 80 percent of the oxygen in the water that passes through them. Teleosts need such a high efficiency because water contains only about 1/30th of the oxygen of air.

Most lungs are much less efficient than gills. Air moves in and out tidally along the same pathway, so there is a little mixing of fresh and stale air. Only 20 percent or so of the oxygen is removed from inhaled air, but since there is plenty of oxygen available, this small percentage is not a problem. However, birds use oxygen at much faster rates than other land vertebrates. They have a crosscurrent system; air moves perpendicular (at 90 degrees) to the blood flow. This system in birds is almost as efficient as fish gills.

Saving water

For oxygen to diffuse, respiratory surfaces must be moist. That requirement can cause problems for land animals. Amphibians must remain close to pools or streams for that reason. Most amphibians that live in arid areas, such as spadefoot toads, spend much of their lives in a state of inactivity called estivation. An exception to the rule that moist-skinned amphibians living in arid areas must remain inactive for most of the time is the Gran Chaco monkey tree frog: it lives in dry areas,

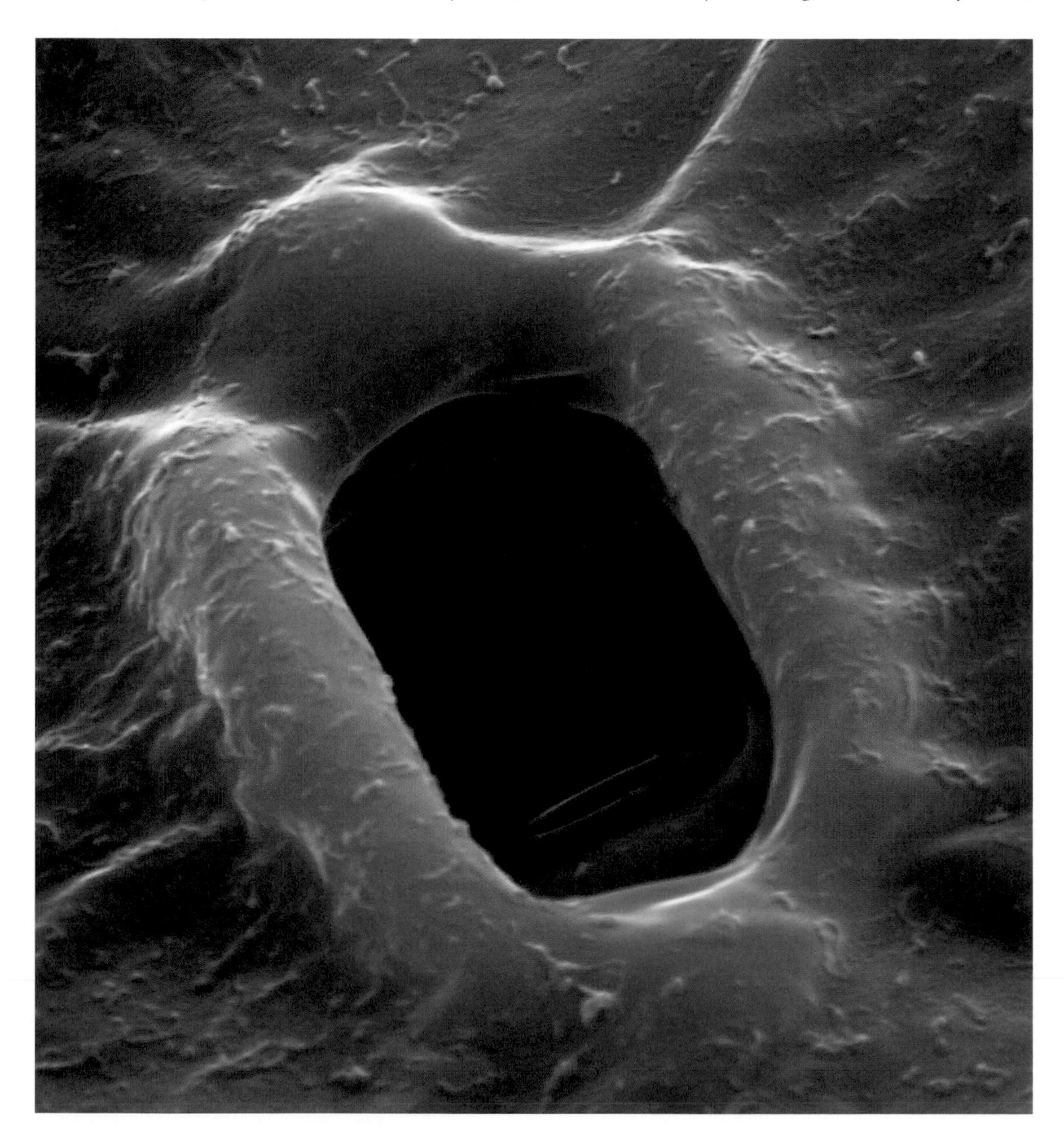

▶ *The stomata, or air pores, of certain desert plants, such as aloe vera, open only at night. Thus they minimize the loss of water through the stomata during the hot day. In these plants, called CAM plants, the carbon dioxide collected at night is stored as an acid, before being broken and used to form sugars during the day (magnified 2,000 times).*

but secretes a waxy substance that it wipes over its skin to resist water loss.

Land animals such as reptiles and insects have waterproof skin and have the ability to shut off their respiratory surfaces. Mammals and birds breathe relatively rapidly, so potentially more water can be lost. They solve this problem with a series of bony or cartilaginous ridges called nasal turbinates. These ridges run from the back of the nose to the epiglottis and are richly supplied with blood vessels. The nasal turbinates warm incoming air and moisten it to prevent the airways from drying. They then recapture water vapor as it is breathed out.

▲ *Atlantic spotted dolphins "porpoise" when swimming fast. Leaping out of the water allows them to breathe efficiently without slowing down.*

IN FOCUS

How plants save water

Plants that live in dry areas fix carbon dioxide into sugars in a different way from other plants to help them save water. Most plants use a process called C3 photosynthesis, so named because carbon dioxide is first fixed into a molecule with three carbon atoms. These plants keep their stomata (air pores) open all day long. Plants in dry areas use a different process, called C4 photosynthesis; they first incorporate carbon dioxide into a four-carbon molecule. This process is much quicker than C3 photosynthesis, so long as the weather is hot and bright, and the stomata can be closed for longer periods in the day. Plants such as corn are C4 photosynthesizers. Desert species such as cacti are called CAM, or crassulacean acid metabolism, plants. Cacti stomata open only at night when it is cool, so little water is lost. Carbon is stored in an acid, which is broken down for incorporation into sugars during the day.

Efficient ventilation at speed

Many mammals do not actively breathe when they are running—the power of movement does the job for them. Once a horse breaks into a gallop, its breathing and stride patterns coincide closely. As it extends its forelegs, the rib cage and viscera, or internal organs, shift forward, pulling the diaphragm and causing inhalation. When the forelegs touch down, the viscera shift back and the rib cage is compressed, forcing air from the lungs.

Swimming birds and mammals also optimize breathing efficiency. Penguins and dolphins often "porpoise" at high speed: they leap through the air as they swim. Swimming part-submerged at the surface uses more energy than swimming underwater. Porpoising allows the animals to breathe without slowing down and minimizes the energy used for swimming.

JAMES MARTIN

FURTHER READING AND RESEARCH

Kardong, Kenneth V. 2005. *Vertebrates: Comparative Anatomy, Function, Evolution*. McGraw-Hill Education: New York.

Raven, Peter H., George B. Johnson, Susan R. Singer, and Jonathan B. Losos. 2004. *Biology*. McGraw-Hill Science: New York.

Rhinoceros

ORDER: Perissodactyla FAMILY: Rhinocerotidae
GENERA: *Dicerorhinus, Rhinoceros, Diceros,* and *Ceratotherium*

Of the five species of rhinoceroses—or rhinos—living today, two inhabit the open savannas and thornbush of sub-Saharan Africa (the white rhinoceros and the black rhinoceros) and three live in the forests of southern and Southeast Asia (Indian, Javan, and Sumatran rhinoceroses). Unfortunately, these unique prehistoric-looking mammals have become increasingly rare in recent decades owing to poaching and loss of habitat. There are now fewer than 12,500 living rhinos of all five species; four out of the five species are listed as critically endangered by the International Union for the Conservation of Nature; the fifth, the Indian rhinoceros, is classified as endangered.

Anatomy and taxonomy

Most organisms are related to one another in some way or other. Closely related species share a relatively recent common ancestor and have had less time to evolve different shapes and lifestyles. Other species have separated from their common ancestor in the more distant past and have had vast amounts of time to evolve into different forms. Because evolution by natural selection works by making small changes to existing anatomical features, related animals always share basic anatomical features. The more closely related the species, the more anatomical features they have in common.

- **Animals** All animals are multicellular and depend on eating other organisms for food. Unlike other multicellular organisms, such as plants and fungi, most animals are able to move about and react quickly to outside stimuli.

- **Chordates** At some time in their life cycle, all chordates have a stiff supporting rod, called a notochord, running along the back of their body.

- **Vertebrates** All vertebrates develop a notochord consisting of a spinal cord encased by a backbone made up of bones called vertebrae. All vertebrates have a skull that is made of either bone or cartilage (a tough, flexible connective tissue) surrounding a brain.

- **Mammals** Mammals are unique among the vertebrates in having milk glands with which the females provide their

▼ *This family tree shows that there are four genera of rhinoceroses, which belong to the family Rhinocerotidae. Only living groups and species are shown in this tree.*

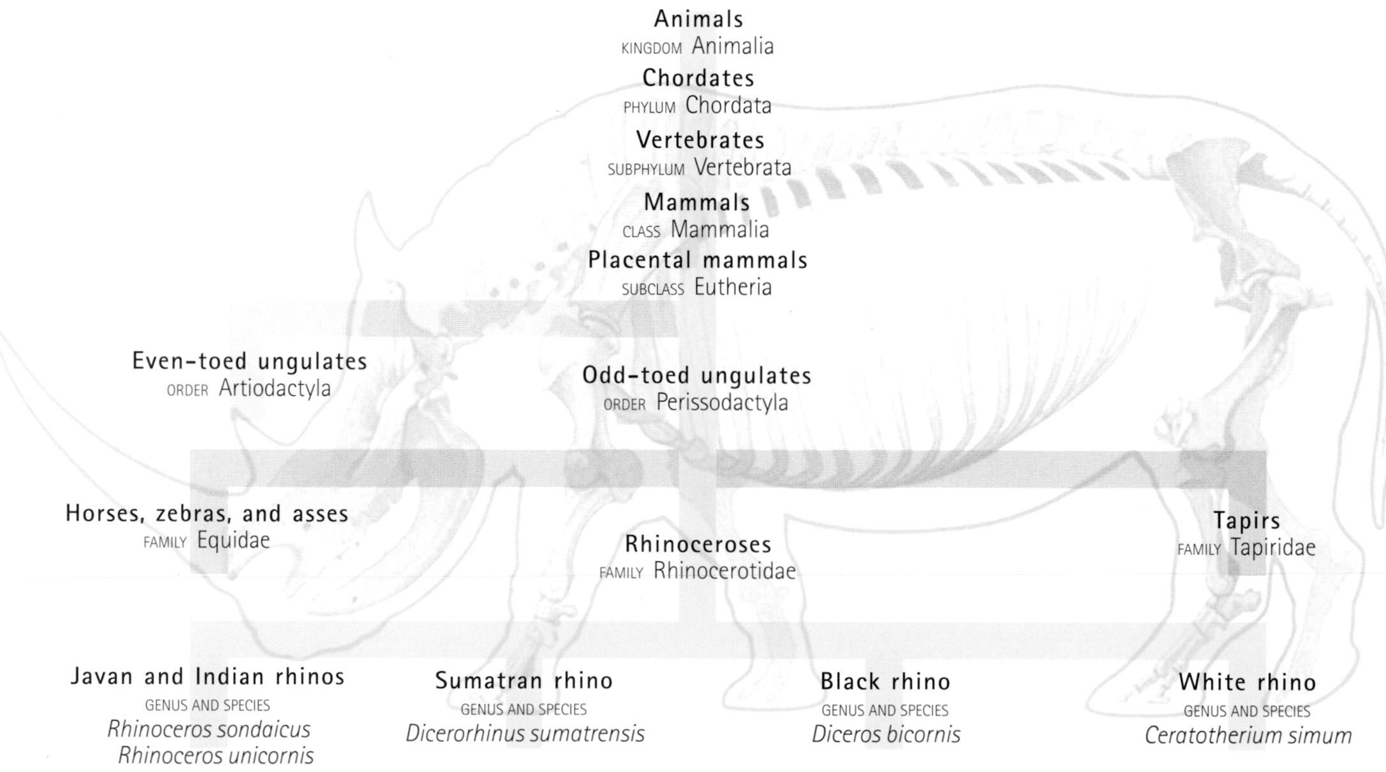

young with milk. Unlike all other vertebrates, mammals have fur, a single lower jawbone that hinges directly onto the skull, and red blood cells that lack a nucleus. Like birds, mammals are warm-blooded and have a four-chamber heart.

▲ *Of the five living species of rhinoceroses, two species (the black and white rhinoceroses) inhabit Africa, and the other three (Indian, Javan, and Sumatran rhinoceroses) live in Asia. The white rhino, above, is the largest species of rhinoceros and lives in the savanna, where it grazes on grass.*

• **Placental mammals** These mammals develop inside the mother's uterus (womb) where they receive nourishment and oxygen from their mother through the placenta, an organ that develops during pregnancy.

• **Odd-toed ungulates** The rhinos share the order Perissodactyla with the horse family and the tapir family. These medium to large grazing and browsing animals are characterized by well-developed third digits that bear most or all of the body weight.

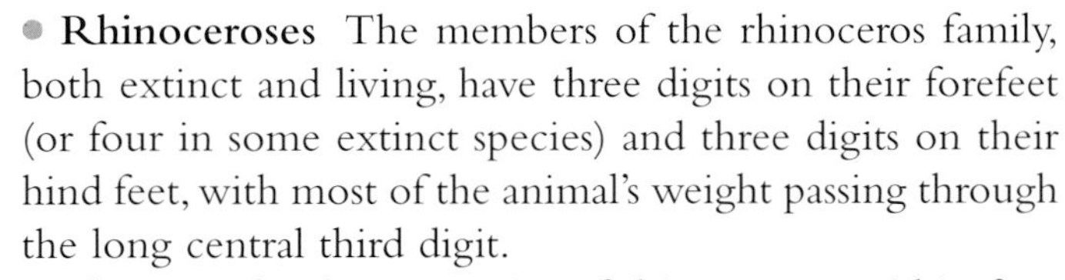

• **Rhinoceroses** The members of the rhinoceros family, both extinct and living, have three digits on their forefeet (or four in some extinct species) and three digits on their hind feet, with most of the animal's weight passing through the long central third digit.

There are five living species of rhinoceroses within four genera. The Sumatran rhinoceros is the smallest living member of the family and has considerably more fur than the other species of rhinos. The Javan rhinoceros and the Indian rhinoceros belong to the same genus; these one-horn rhinos have very thick skin, which unlike that of their African relatives bears a number of loose folds, making them look armor-plated. The black rhinoceros lives in Africa and has a distinctive protruding prehensile (capable of grasping) upper lip, or proboscis. The largest of all living rhinos is the white rhinoceros, and like the black rhinoceros it lives in Africa and has a long frontal (anterior) horn and a shorter rear (posterior) horn. Unlike the black rhinoceros, the white rhinoceros has a squared upper lip with no trace of a proboscis.

FEATURED SYSTEMS

EXTERNAL ANATOMY Rhinos are large four-legged mammals, with thick, leathery skin and a massive head. Most species have one or two conspicuous horns on the head. *See pages 1034–1037.*

SKELETAL SYSTEM The rhinoceros's skeleton bears great weight, with thick limb bones arranged like vertical columns, so the animal's weight passes along the length of the bones. This arrangement reduces the need for muscular exertion and minimizes the risk of injury to the joints. *See pages 1038–1039.*

MUSCULAR SYSTEM Powerful leg muscles help the rhinoceros maintain its posture when walking. The massive head and neck are supported by strong muscles that enable the horns to be wielded with power. *See pages 1040–1041.*

NERVOUS SYSTEM Rhinoceroses have a relatively small brain for their size. They have poor eyesight but good hearing and an excellent sense of smell. *See page 1042.*

CIRCULATORY AND RESPIRATORY SYSTEMS The large, compact body conserves heat very well, so the animal does not need to generate much heat. A slow heart rate is sufficient to supply body tissues with oxygen. *See page 1043.*

DIGESTIVE AND EXCRETORY SYSTEMS Rhinos eat large amounts of low-quality food. It is digested with the help of microorganisms that live in the rhino's enlarged colon. *See pages 1044–1045.*

REPRODUCTIVE SYSTEM Male rhinos have internal testes and a backward-curving penis. Females have a bicornuate uterus (consisting of two tubes). *See pages 1046–1047.*

External anatomy

CONNECTIONS

COMPARE the posture of a rhino with that of an ***ELEPHANT*** and ***HIPPOPOTAMUS***. These are all huge herbivorous (plant-eating) mammals that have an upright stance and straight legs to carry their great weight with minimal effort.

All species of rhinoceroses have a large head, powerfully built shoulders, and a massive body supported by thick legs. The forelegs, which support all the weight of the rhino's head and massive forequarters, are straight and pillarlike, whereas the hind legs have a distinct heel. Body proportions vary among species. For example, the Sumatran rhino has a short body; the white rhino has a well-developed shoulder hump; and the Javan rhino has a smaller head than the closely related Indian rhino. All species of rhinoceroses have a tail ranging from 20 to 40 inches (50–100 cm) long.

The rhinos are among the largest living land mammals, rivaled in size by only the elephants and common hippopotamus. The largest species of rhino is the white rhinoceros, with males weighing up to 7,940 pounds (3,600 kg). Female white rhinos are only about half the size of the males, usually weighing from 3,100 to 3,750 pounds (1,400–1,700 kg). Other species show little or no difference in size and form between the sexes (sexual

▶ **White rhinoceros**
Weighing up to 7,940 pounds (3,600 kg), the white rhino is the largest living rhinoceros species. It has gray, textured skin; a pronounced hump; and a long head with two horns, pointed ears, and a broad square upper lip.

dimorphism). The other species living in Africa, the black rhino, is less than half the size of the male white rhino, weighing from 1,760 to 3,100 pounds (800–1,400 kg). The smallest rhino is the Sumatran rhino, weighing about 2,200 pounds (1,000 kg). The other two Asian rhinos (Indian and Javan rhinoceroses) range in weight from 3,300 to 4,850 pounds (1,500–2,200 kg).

Head and horns

Perhaps the rhinos' most striking feature is their horns. The two African species bear two horns on the snout, a large anterior (frontal) horn above the nostrils, and a smaller posterior (rear) horn above the eyes. The anterior horn of white and black rhinos

▲ *The white rhino has a large frontal horn and a smaller rear horn. The birds on the rhino's head are oxpeckers, which keep its thick skin free of parasitic ticks.*

COMPARATIVE ANATOMY

Horns

Unlike the horns of hoofed animals such as deer, antelopes, and cattle, the horns of rhinoceroses are not made of bone. Rhino horns are made of compressed keratin, the same fibrous protein found in hair, fingernails, and skin, which derives from the dermal, or skin, tissues. Rhino horns are connected to the skull by thick connective fibrous tissue. Although rhino horns differ in form and origin from the horns of other hoofed mammals, they serve a similar purpose. Male rhinos use them when fighting over access to females. Duels between male black rhinos are sometimes fatal. Female rhinos use their well-developed horns to defend their young from big predators, such as lions and tigers.

averages 20 to 24 inches (50–60 cm) long, but may grow to 4 feet (136 cm) in black rhinos and exceed 5 feet (150 cm) in white rhinos. Female black rhinos usually have larger horns than the males. The Asian rhino species have much smaller horns than their African cousins. The Sumatran rhino, like the African species, has both anterior and posterior horns, although the horns are much smaller, the record length being only 15 inches (38 cm). The Indian and Javan rhinos have only a single frontal horn, with a length ranging from about 6 inches (15 cm) in the Javan rhino to 21 inches (53 cm) in the Indian rhino. Both male and female rhinos have horns, although some female Javan rhinos may lack them.

Rhinos have small eyes in relation to their body size. The eyes are positioned on the sides of the head. This arrangement gives them a wide field of vision. The nostrils are large and particularly pronounced in the white and black rhinos. All rhinos have large upright ears that can be moved to focus on sound sources. Ear shape differs between species. For example, white rhinos have pointed ears fringed with sparse bristles, whereas black rhinos have rounded ears with hairy edges. Black rhinos, Indian rhinos, and Javan rhinos all have a characteristic prehensile proboscis on the upper lip, which they use for helping draw twigs and leaves into the mouth. The white rhinoceros, however, eats only grass, and instead of a proboscis has a squared upper lip that is

▲ Indian rhinoceros

The Indian rhino has a single blunt horn, situated over the nostrils, that can be up to 21 inches (53 cm) long. Its upper lip is prehensile (can grasp).

▲ Sumatran rhinoceros

Like the African rhinos, the Sumatran species has two horns. However, the Sumatran rhino's horns are shorter, the front horn growing to 15 inches (38 cm).

EVOLUTION

Prehistoric rhinoceroses

The oldest fossil rhinoceros was found in rocks originating from the middle Eocene period (about 45 million years ago) in North America. In prehistoric times, a great diversity of mostly hornless rhinos of various shapes and sizes lived in North America and Europe, as well as Africa and Asia. Most varieties died out owing to drastic climate changes in the late Miocene period (about 10 million years ago). Two species that survived this period of extinction into the Pleistocene (5 million to 1.6 million years ago) were the Eurasian woolly rhinoceros (*Coelodonta*) and the elephant-size *Elasmotherium* with its giant frontal horn. Perhaps the largest land mammal to have ever lived was the hornless rhinoceros *Indricotherium*, or *Baluchitherium*, which weighed up to 33 tons (30 metric tons), five times the weight of a modern African elephant.

▼ Extinct rhinoceroses

Indricotherium *is the largest known land mammal to have ever lived. The giant frontal horn of* Elasmotherium *could grow to almost 6 feet (1.8 m) in length. Cave paintings from 30,000 years ago in France show depictions of* Coelodonta.

▲ **Indian and Sumatran rhinoceroses**
The Indian rhino is from 10 to 12.5 feet (3–3.8m) in length. It has distinctive armorlike skin folds. The Sumatran rhino is the smallest species of rhino, 8 to 10 feet (2.5–3 m) long.

perfectly suited for cropping grass. The white rhino has a relatively long head that is held lower than that of the black rhino; it can easily reach short vegetation. The white rhino is the largest strictly grazing animal in the world.

Tough skin

Rhinoceroses have very thick skin that can be more than 1 inch (2.5 cm) thick, protecting the animal from the attacks of predators and also other rhinoceroses. Both species of Asian one-horn rhinos (Indian and Javan rhinos) have conspicuous skin folds, giving these animals the appearance of wearing medieval armor. The folds are most pronounced in male Indian rhinos. The Sumatran rhino also has skin folds, although they are less striking than those of the one-horn rhinos. Neither of the African rhinos has such skin folds. In Indian rhinos, the appearance of armor plating is enhanced by knobby bumps, or tubercles, resembling rivets, on the lower shoulders and haunches and legs. The skin of the closely related Javan rhino is covered with small scalelike disks.

The Sumatran rhino is covered with reddish-brown fur that turns black with age, whereas all other rhinos are practically naked apart from eyelashes, ear fringes, and tail brushes. Rhinos come in various shades of gray-brown, though they are often colored by the mud in which they wallow. The one-horn rhinos sometimes have shades of pink around the skin folds. Despite its name, the white rhino is not white—it is only a little lighter than the black rhino, which varies from dark yellow brown to dark brown or gray. The white rhino's name most likely comes from a corruption of the Afrikaans word *wijde*, meaning "wide," which is probably a reference to this animal's broad mouth and body size.

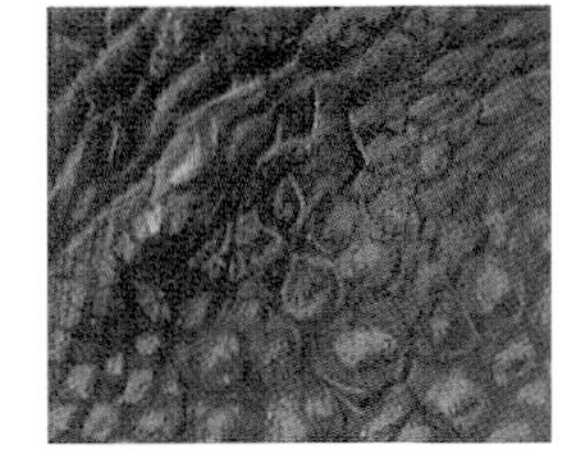

Javan rhinoceros

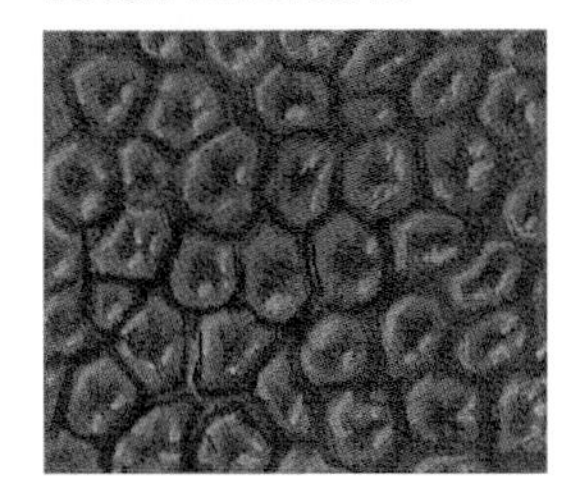

▲ **SKIN SURFACE**
The Indian rhino's skin has wartlike bumps on the shoulders and upper legs, whereas the Javan rhino's skin has a scaly-looking mosaic pattern all over. Both species have hairless skin.

▼ **Black rhinoceros**
The black rhino has two horns; the larger frontal horn can grow to 4 feet (1.4 m) long. The mouth is narrow—compared with that of the white rhino—and the upper lip is prehensile and used for pulling twigs and vegetation into the mouth.

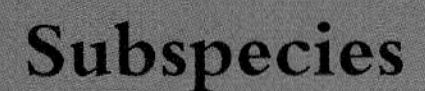

Subspecies

Populations of different subspecies (local forms) each have a distinct genetic makeup as well as slightly different physical characteristics. Some rhino species can be divided into two or more subspecies. For example, the rare Sumatran rhino has three subspecies: an extinct one that used to live in India, Bangladesh, and Myanmar; another from Borneo (only 60 survivors); and the third from Sumatra, Peninsular Malaysia, and Thailand. There are up to seven subspecies of black rhino (though the exact number is disputed), some of which have already become extinct; and two subspecies of white rhino.

Skeletal system

> **CONNECTIONS**
>
> **COMPARE** the rhino's skeleton with that of an ***ELEPHANT*** and a ***ZEBRA***. Elephants are graviportal: they have a skeleton with heavy weight-bearing characteristics. Lighter running animals, such as the zebra, are cursorial: they have features for speed, such as long limb bones. Rhinos are mediportal: they have less extreme features for weight bearing than elephants.

Like all mammals, a rhinoceros has a skeleton divided into four major sections: the skull, the backbone, the ribs, and the limb bones. The rhino's skeleton shows a number of features suited to bearing the animal's heavy body weight efficiently and also for allowing the animal to move quickly when necessary. Heavy animals such as rhinos have thicker, more bulky bones than lighter species.

The backbone

In mammals, the backbone (spine, or vertebral column) can be divided into five sections: cervical (neck), thoracic (upper back), lumbar (lower back), sacral (just above the tail), and caudal (tail) vertebrae. The thoracic and lumbar sections of a rhinoceros's backbone need to be able to support the considerable weight of the digestive system and other internal organs, as well as all the surrounding muscle. As in other large mammals, these sections of the rhinoceros's backbone are slightly arched like a bridge in order to provide extra strength. The thoracic vertebrae closest to the shoulder have particularly long dorsal (upper-side) processes. These processes are most notable in the white rhinoceros and act as anchorage points for strong ligaments and large muscles that extend to the back of the skull and help support the animal's massive head.

Sturdy limbs

The bones in a rhinoceros's limbs are specially designed and arranged to allow the rhinoceros to stand and move with little effort and with limited risk of injury. Like any rod-shape objects, long bones are better at withstanding compressive forces along their length than bending forces acting from the side. For this reason, rhinoceros bones have a straight shaft and are aligned vertically, so the bodyweight

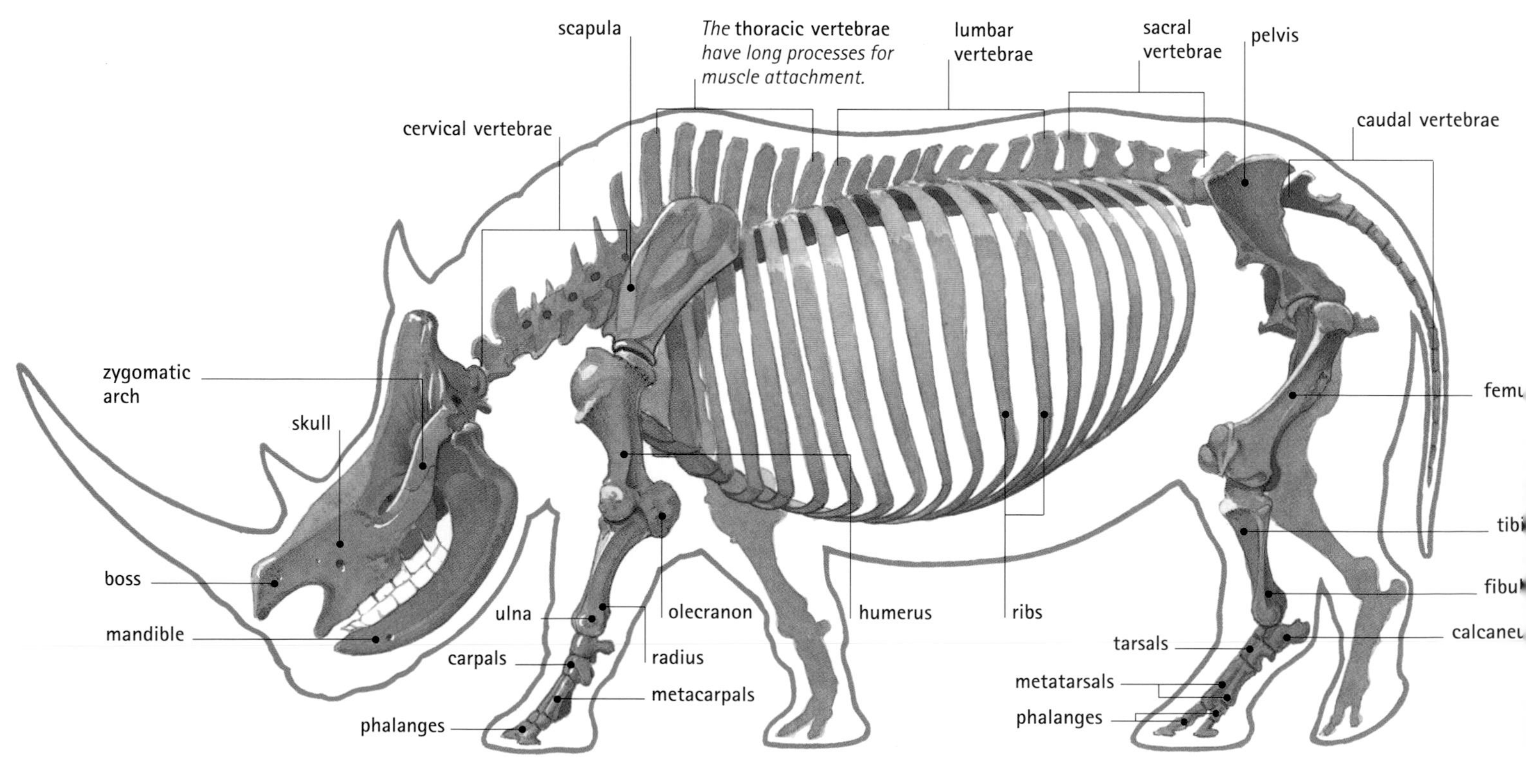

▼ **White rhinoceros**

The rhino has a massive skull with a bony mound, or boss, on top of which the frontal horn grows. The strong spine supports the animal's great weight, as do the sturdy vertically arranged bones of the legs. The deep rib cage protects the internal organs.

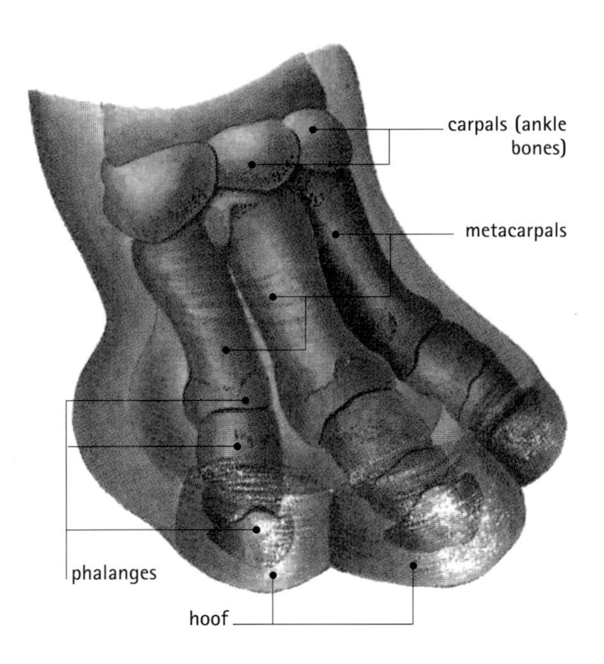

▲ RHINO FOOT

The rhino is unguligrade—it walks on tiptoe. Each foot has three digits composed of sturdy, blocklike phalanges. The digits are surrounded by a horny covering, or hoof, with most of the animal's weight passing through the central long digit.

passes through the length of the bones. The limb bones of the rhinoceros have a number of features that maintain the bones in their vertical alignment.

Unlike that of humans, the rhino's shoulder blade (scapula) is positioned directly above the upper arm bone (humerus) and is not held in place by a collarbone (clavicle), which is absent in the rhinoceros. As in an elephant's limb, the olecranon process (elbow) on the rhino's lower arm bone (radius) is deflected backward, thus allowing the humerus to align vertically with the radius. The heads of the humerus and thighbone (femur) point nearly straight upward, so they can transfer weight directly down the length of the bone. Compare this arrangement with the head of the femur in the human skeleton, in which the head sticks out sideways to connect with the hip bones (pelvis). In the rhinoceros, the radius and ulna of the lower forelegs and the fibula and tibia of the hind legs are well separated. The wrist bones (carpals) and ankle bones (tarsals) of the rhino are short and blocklike in relation to those of smaller, more agile hoofed mammals.

CLOSE-UP

The skull

Rhinoceroses have a long, heavily built skull that curves upward at the posterior end (the occipital region). The skull is characterized by a small braincase, a long anterior section (facial bones), and massive mandibles (lower jaws). The raised occipital region at the back of the skull provides a large attachment area for the powerful neck muscles that lift the massive head. White and black rhinos have large bulbous nasal bones, which provide a strong base for the horns. The rough upper surface of the nasal bones provides a good attachment site for the fibrous connective tissue that joins the horns to the skull.

▼ SKULLS

Rhino skulls have a large bony mound, or boss, on which the frontal horn grows. The long, deep occipital region at the back of the head provides an attachment area for the large muscles necessary to hold up the head.

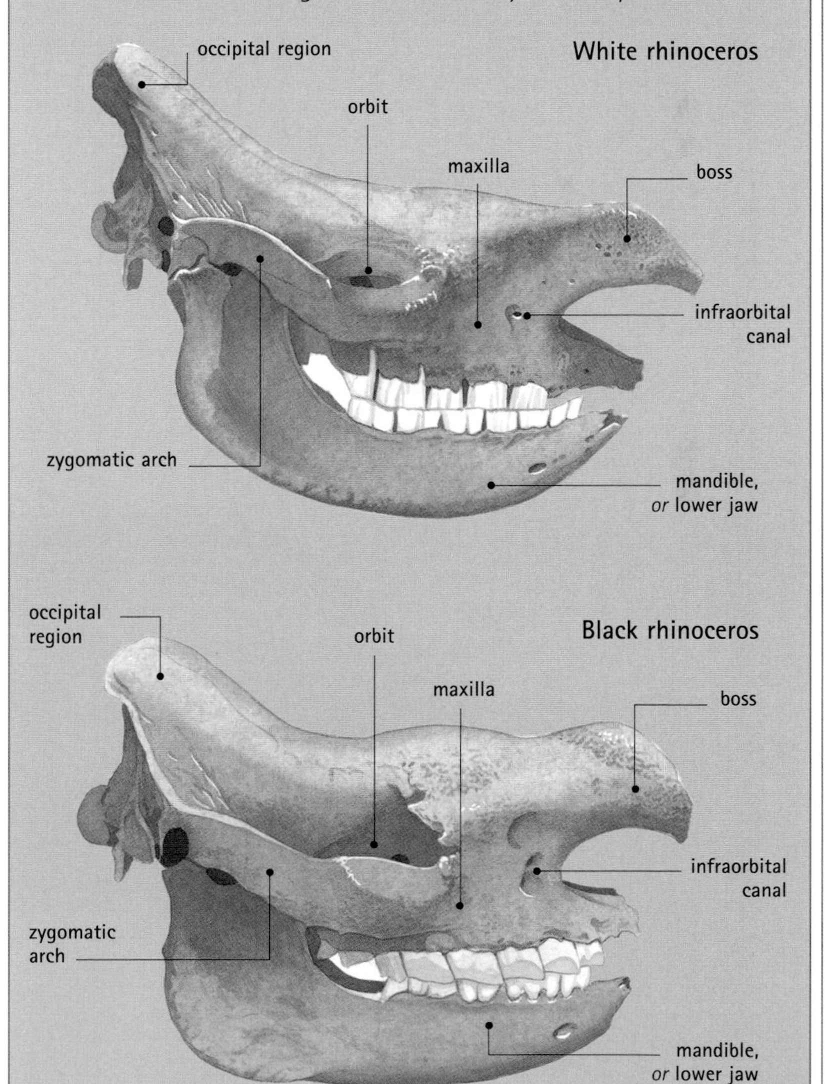

Muscular system

CONNECTIONS

COMPARE the muscular system of a rhino with that of an ***ELEPHANT*** or ***HIPPOPOTAMUS***. All these animals have powerful neck and leg muscles. Lighter ungulates such as horses, antelopes, and deer have a sleeker body shape, reflecting different muscular proportions suited for running.

There are three main types of muscles in vertebrates: skeletal, or striated, muscle; smooth muscle; and cardiac muscle. Skeletal muscles are the large muscles attached to the skeleton and are used for maintaining posture, locomotion, and other body movements. They are also called striated muscle because, when viewed under a microscope, the muscle fibers have distinctive dark and light bands called striations. Smooth muscles are made up of shorter muscle cells, and they are present in many internal structures, such as the esophagus, stomach, intestines, and blood vessels. Smooth muscles are responsible for moving food down the throat during swallowing and then through the intestines. Cardiac muscle is found only in the heart walls and is responsible for pumping blood around the circulatory system.

In most vertebrates, the structure of skeletal muscles is symmetrical on each side of the body. The muscular systems of different mammals are structurally similar and mostly made up of the same muscles. For example, forelimb muscles—such as the biceps, which flex the forelegs; and the triceps, which extend the forelegs—occur in both rhino forelegs and the human arm. Many skeletal muscles, such as the triceps and biceps, act as antagonistic pairs, working against each other in opposite directions. Because rhinos have a different posture from humans, the relative shapes and sizes of their muscles also differ. The foreleg extensor muscle, the triceps, which maintains the rhino's upright posture when walking, is large and strong. The muscles of the neck, such as the trapezius and rhomboideus, are large in rhinos and so are able to maintain the heavy

▼ **White rhinoceros**

Rhinos have very large neck muscles to hold up the heavy head. These muscles help to produce the animal's shoulder hump. The legs also have powerful muscles to maintain a tiptoe stance.

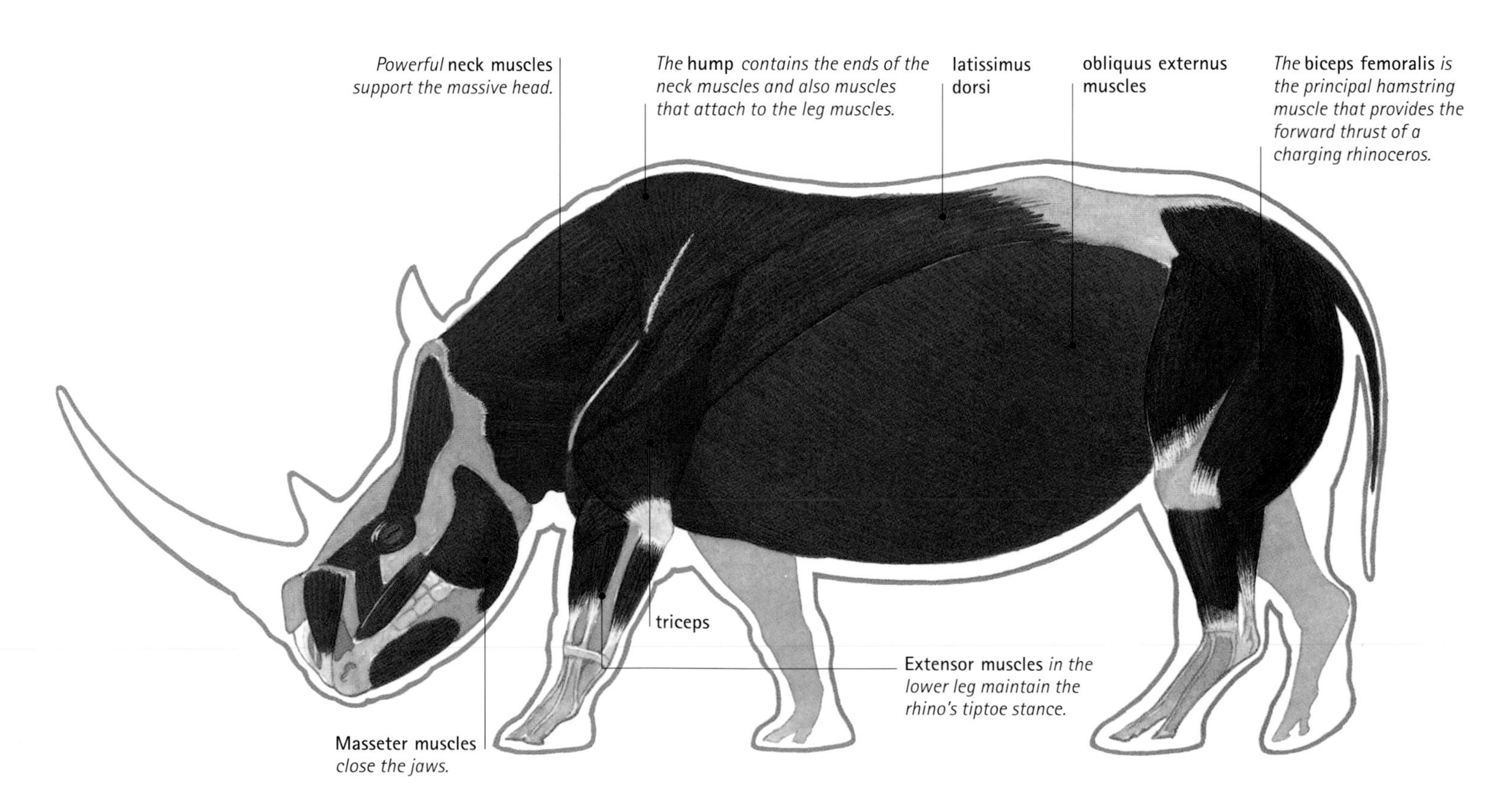

▲ *The white rhino has a distinct hump on its back. The hump is where of the ends of large muscles that support the head attach to the long projections of the thoracic vertebrae of the backbone.*

head in an upright position. These powerful muscles are rooted in the white rhino's conspicuous hump. The hump also contains muscles that help power the forelegs.

To maintain the rhino's unguligrade stance, the lower legs have powerful wrist and ankle extensor muscles, such as the flexor carpus (forearm muscle) in the forelegs and the gastrocnemius (calf muscle) in the hind legs. The largest muscles in the rhino's legs are the principal hamstring muscles, or biceps femoris. The hamstring muscles flex the hind legs and are used to provide much of the forward thrust of a running rhinoceros.

There are only 16 main muscle groups around the body wall, and these include some of the largest muscles in the rhinoceros's body, such as the latissimus dorsi on the flanks behind the shoulders (corresponding with the upper, outer back muscles in humans). In addition to maintaining posture, the muscles of the body wall help support and hold in the internal organs of the rhinoceros, such as the tightly packed intestines, kidneys, and liver.

CLOSE-UP

Facial and head muscles

It is not just the large muscles that are important. Like most mammals, rhinos have a large number of small muscles in the head that control the movement of the jaws, lips, and eyes. Unlike humans, rhinos have well-developed auricularis muscles for moving the ears and nostril-dilator muscles for flaring the nostrils. Several strong muscles in the upper lip serve the prehensile lips of the black rhino, allowing the lips to reach and grasp. Rhinos also have powerful chewing muscles, the masseters, allowing them to grind rough vegetation.

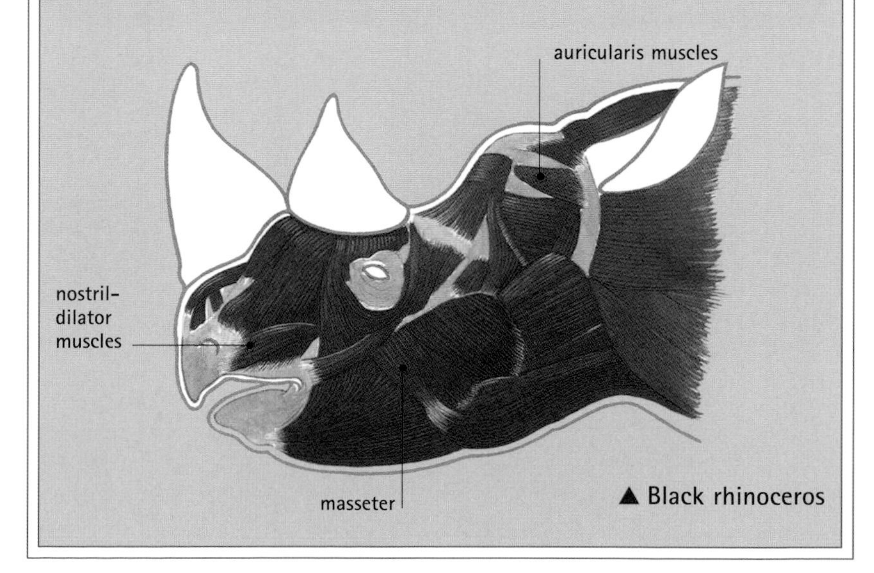

▲ Black rhinoceros

Nervous system

Like all mammals, the nervous system of a rhinoceros can be divided into two parts: the central nervous system (CNS), which is made up of the brain and the spinal cord; and the peripheral nervous system (PNS), which includes all the other nerves that lead to and from the brain and spinal cord. In all vertebrates, the spinal cord passes through the middle of the backbone (vertebral column).

COMPARATIVE ANATOMY

Small-brained animals

Rhinos have a smaller brain than expected for their size: it usually weighs less than 1 pound (0.45 kg). Compare this with the 14-pound (6.5 kg) brain of the African elephant. The small size of a rhino's brain may be partly due to the solitary nature of this animal, which contrasts with the complex social behaviors of elephants.

Types of neurons

The nervous system is made up of cells called neurons that transmit electrical signals. There are three types of neurons: sensory neurons, which conduct electrical signals from sense organs to the CNS; motor neurons, which carry signals to the muscles or to an effector organ such as a gland to initiate a response; and interneurons, which connect the sensory and motor neurons. Nerves are bundles of neurons with small blood vessels that bring nutrients and oxygen to the neurons.

Like all vertebrates, rhinoceroses have a pair of large nerves leaving each side of the spinal cord at regular intervals. These nerves quickly divide into many branches, leading to specific parts of the body. For example, the nerves leaving the thoracic region of the spine carry sensory and motor neurons to and from organs such as the heart, lungs, stomach, liver, pancreas, and kidneys as well as to and from many of the muscles in the neck and chest.

Rhinos gather information from the world around them by using their sight, hearing, smell, touch, and taste. They have very poor eyesight but possess a very keen sense of smell and can detect chemical signals called pheromones left by other rhinoceroses. Rhinoceroses can also smell approaching predators from a distance.

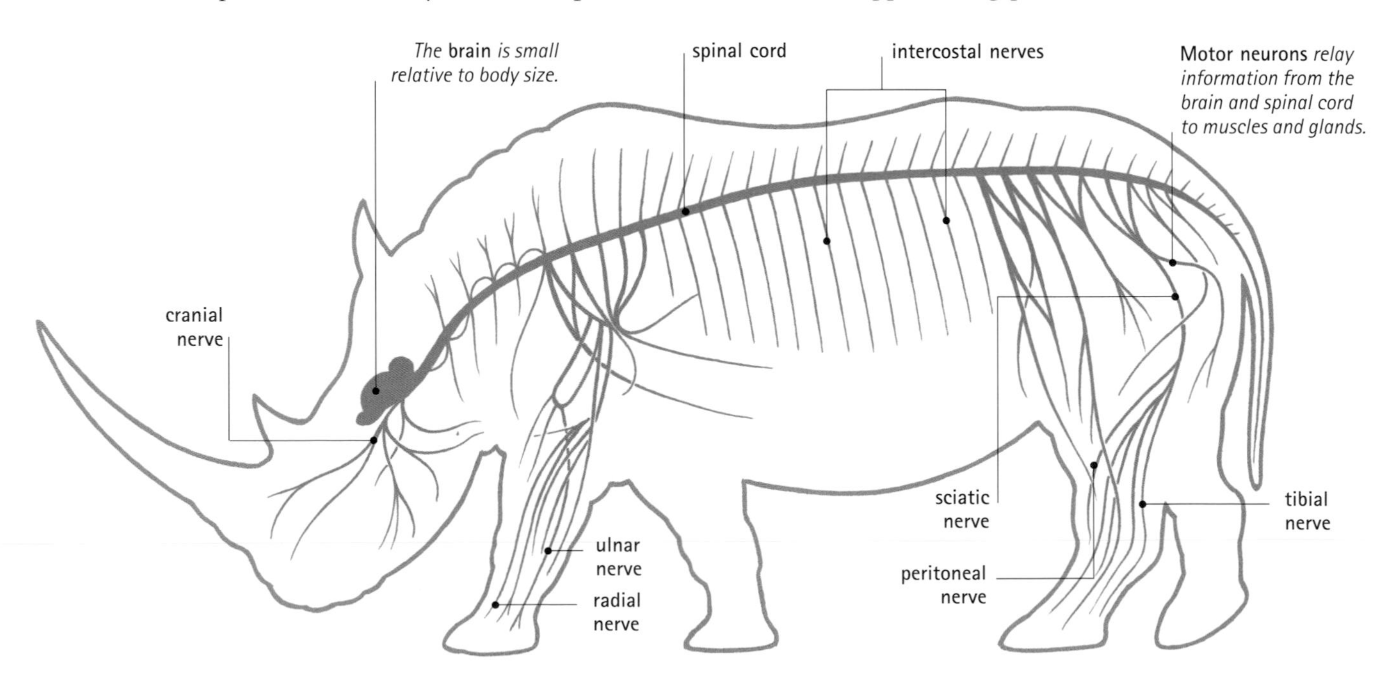

▼ White rhinoceros *Rhinoceroses have a relatively small brain for their size. Their vision is poor, but they have good hearing and an excellent sense of smell.*

Circulatory and respiratory systems

Rhinoceroses have the same basic plan as other mammals for their circulatory and respiratory systems.

In common with other mammals, rhinos have a four-chamber heart with two ventricles (upper chambers) and two atria (lower chambers). Arteries carry oxygenated blood, whereas veins carry deoxygenated blood. Deoxygenated blood from the rhino's body passes into the right atrium via a pair of thick veins called the venae cavae. Contraction of the right atrium forces blood through a one-way valve into a larger chamber, the right ventricle, which then squeezes the blood through the small blood vessels (capillaries) passing through the lungs.

Oxygenated blood returning from the lungs then passes into the left atrium, which pushes the blood through another one-way valve into the most powerful section of the heart, the left ventricle. The left ventricle needs to be powerful to pump the oxygenated blood through the main artery (the aorta) into the rest of the rhino's body. Rhinoceroses' arteries are large and heavy and are supported by ridges of connective tissue or muscles; and the veins are wide, with especially thick walls that keep them from collapsing. The artery walls also need to be thick to withstand a high blood pressure.

IN FOCUS

Heart rate and energy efficiency

Like other large mammals, rhinos have a slow heart rate—for example, 30 to 40 beats per minute in the white rhino. A rhinoceros can afford to have a comparatively slow heart rate and hence a relatively slow rate of oxygen delivery to the body because per pound of body weight it needs less energy than smaller mammals. The rhino's large compact body conserves heat very well, so it does not need to generate as much heat—which requires oxygen—as smaller mammals.

Respiratory system

In rhinos, like all mammals, air passes down a windpipe (trachea) and into two lungs via the bronchial tubes. Most mammals have a space between the lungs and the chest wall (pleural cavity). Raising the ribs and lowering the diaphragm muscle increases the volume of the pleural cavity, creating a negative pressure, which causes the lungs to inflate with air.

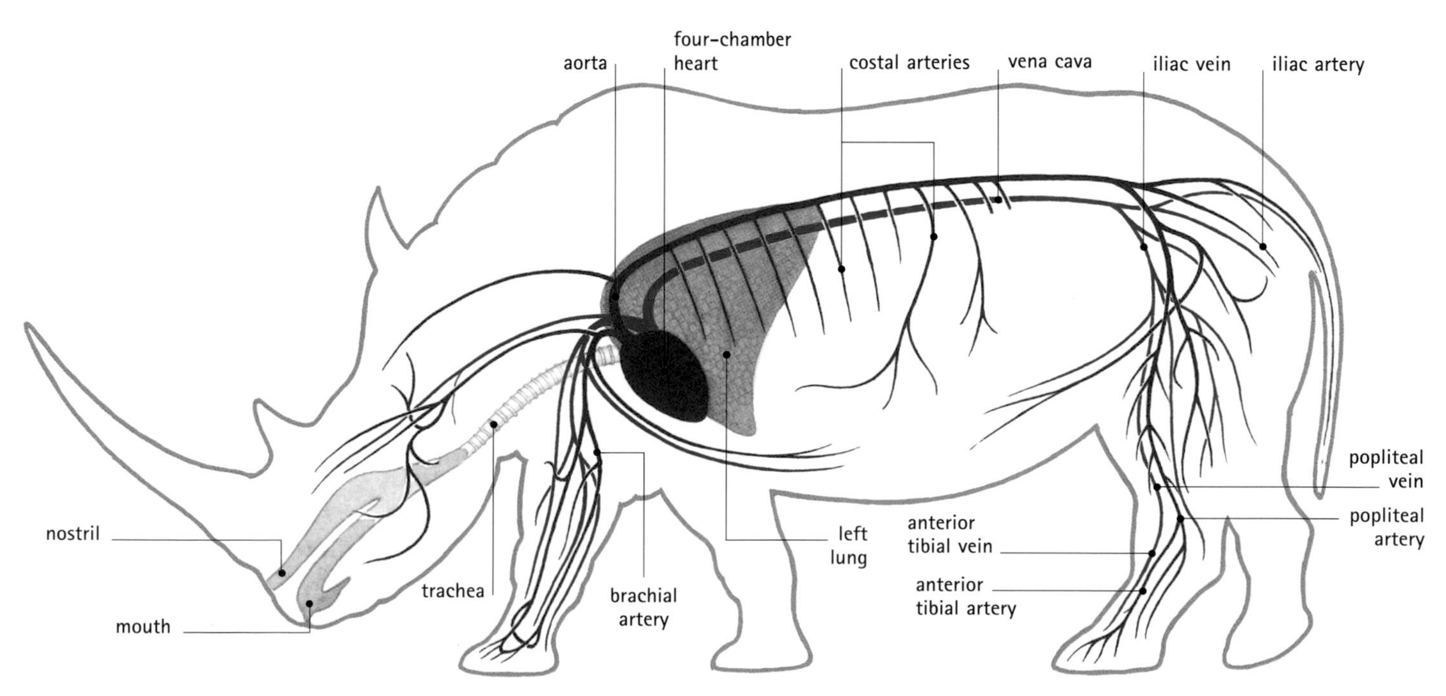

▼ White rhinoceros *A four-chamber heart pumps blood to the lungs, where it picks up inhaled oxygen. This oxygenated blood returns to the heart to be pumped around the body to all tissues.*

Digestive and excretory systems

CONNECTIONS

COMPARE the structure of a rhino's stomach with that of ruminants such as ***RED DEER***, which have a complex multichamber stomach for the fermentation of cellulose. In rhinos, the cellulose is fermented farther along the digestive tract in the colon and in a chamber called the cecum.

The white rhino eats only grass, but other species of rhinos eat a wide range of vegetation, including grass, leaves, fruit, and woody material such as twigs and bark. This type of food is not very nutritious and is difficult to digest; rhinos can survive on this diet because of their large size and low energy requirements. Being large, rhinos can accommodate the long and voluminous digestive system needed to digest and absorb coarse vegetation.

Teeth and salivary glands

The digestion process starts in the mouth, where vegetation is chewed up and mixed with saliva. Rhinoceroses chew with between 24 and 28 great molars and premolars, depending on the species. The first premolars, closest to the front of the mouth, are small; but the rest are large and resemble the molars. These huge cheek teeth are covered with enamel ridges that help grind up vegetation. The grass-eating white rhino has high-crowned cheek teeth—extending far above the gum line—whereas those of the other species are low-crowned. Unlike their African relatives, Asian rhinos also have large lower incisors. In Indian rhinos, these tusklike teeth are used for stabbing during fighting.

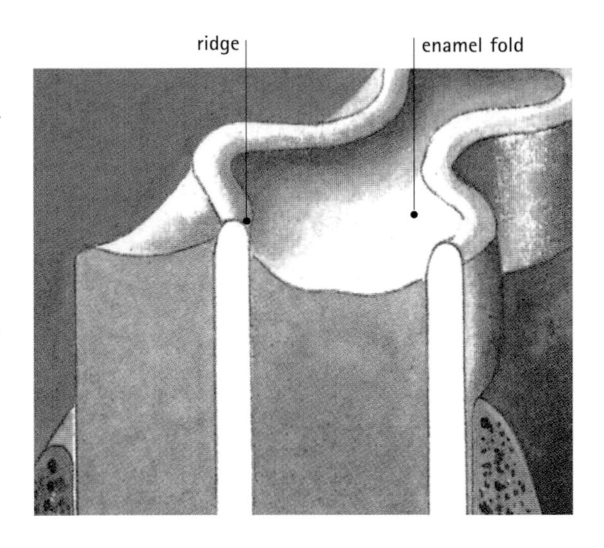

▲ **CROSS SECTION OF TOOTH**
The rhino's cheek teeth have folds in the enamel surface, which form ridges as the teeth become worn down with age. These folds and ridges are effective for grinding and shredding vegetation.

Rhinoceroses have well-developed salivary glands and also mucus glands in the esophagus. These moisten dry vegetation, allowing it to move easily down to the stomach. The rhino's digestive system is fairly simple compared with that of other large plant eaters. Rhinoceroses have a relatively small and simple stomach.

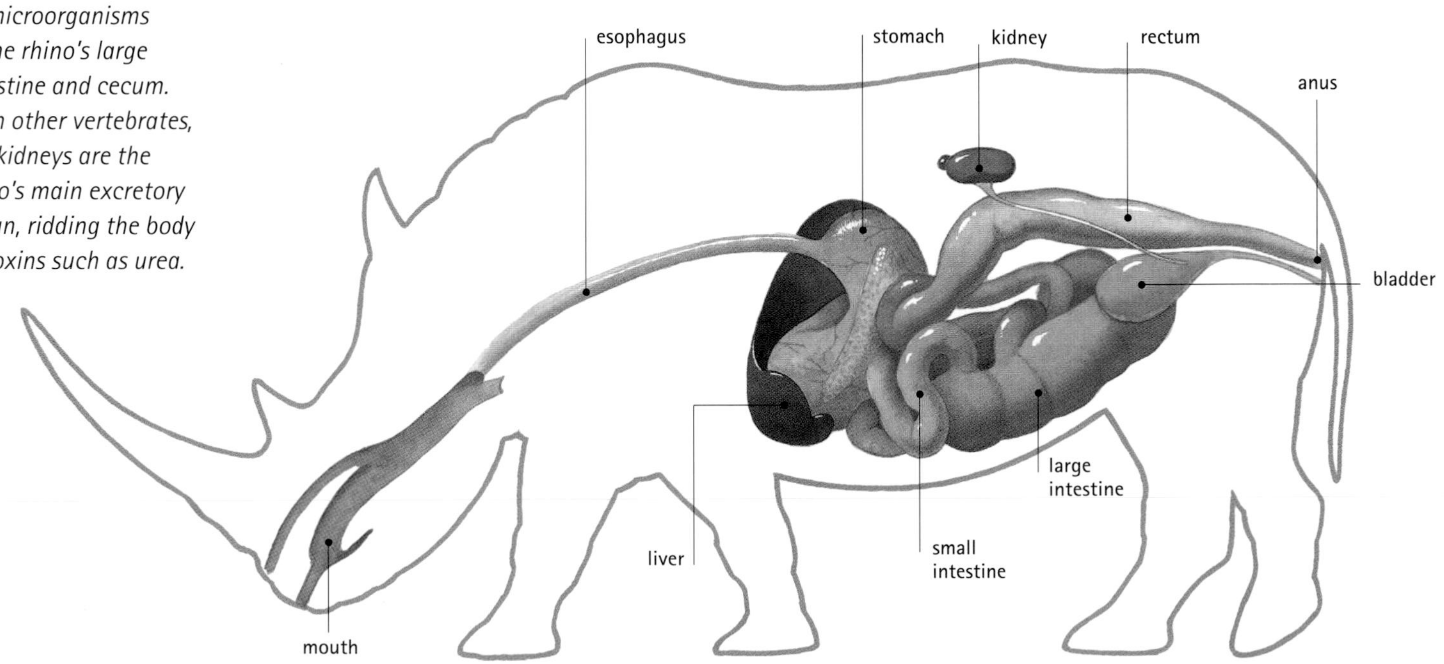

▼ **White rhinoceros**
The rhino's long digestive system is suited to a diet of coarse vegetation. The tough carbohydrate cellulose, found in plant cells, is broken down by microorganisms in the rhino's large intestine and cecum. As in other vertebrates, the kidneys are the rhino's main excretory organ, ridding the body of toxins such as urea.

▲ MOUTH
Black rhinoceros
The black rhino, also called the hook-lipped rhino, has a prehensile (gripping) upper lip, which is suited to drawing the leaves and twigs of bushes into the mouth.

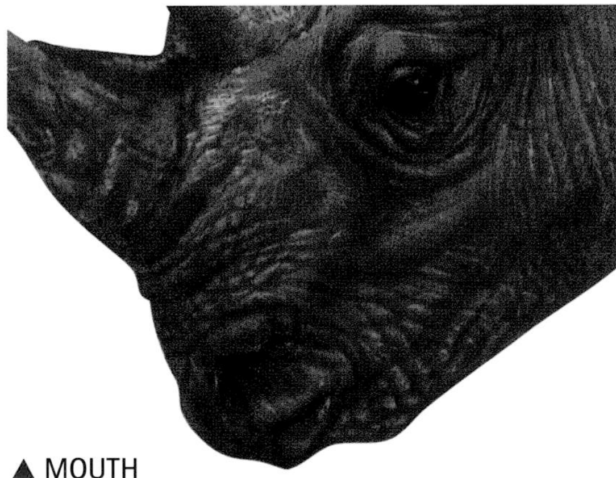

▲ MOUTH
White rhinoceros
The white rhino, also called the square-lipped rhino, has a wide mouth. The upper lip is not prehensile. This species eats only grass.

Symbiotic microorganisms

Some types of plant-eating animals, such as the ruminants, which include cattle, antelopes, deer, and giraffes, have a multichamber stomach in which the indigestible carbohydrate cellulose (the main constituent of plant cell walls) is broken down by microorganisms. Rhinos also need symbiotic microorganisms to ferment and digest cellulose, but instead of occurring in the stomach this process occurs in the rhino's colon (the major part of the large intestine) and in a chamber called the cecum. The cecum lies at the junction of the small and large intestines. The nutrients that are released from fermentation of cellulose are absorbed directly through the wall of the colon, which has many blood vessels to transport the nutrients around the body.

The indigestible remainder of the food passes from the upper to the lower part of the large intestine, where water absorption occurs; and then to the rectum, where the feces are stored until ejected as droppings. White rhinos defecate five or six times a day, and some rhinos, such as the black rhino, create conspicuous communal dung piles up to 3 feet (1 m) high. These dung piles are marked with pheromones that identify the sex and reproductive status of local rhinos.

Excretory system

Unlike smaller mammals, rhinos have kidneys made up of separate cone-shape lobes (up to 80 in Indian rhinos), presumably for maintaining structure and shape. Kidneys are the principal excretory organ, and they are responsible for removing toxins such as urea from the blood and regulating blood-salt concentration. The kidneys filter blood through small bundles of capillaries called glomeruli. Rhinoceroses have relatively small glomeruli, but these are very abundant—up to 16 million in each kidney of the Indian rhino, for example. Except for their size, the other organs involved in digestion and excretion in rhinos, such as the liver and pancreas, are typical of mammals.

▼ LIVER
CROSS SECTION
In rhinos, as in all mammals, the liver has many roles, including the production of bile to aid the digestion of fats, and the removal of poisonous substances in the blood (detoxification).

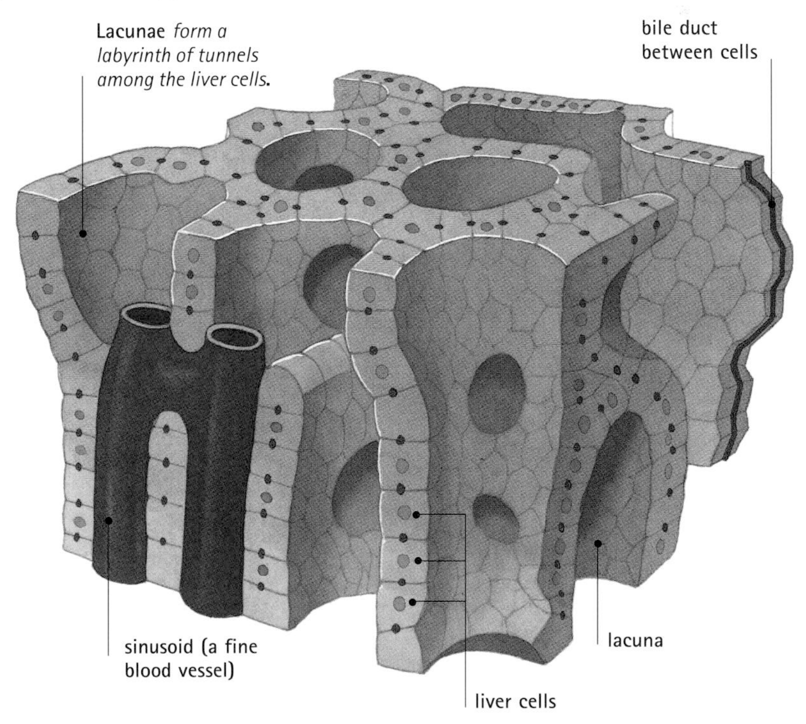

Reproductive system

CONNECTIONS

COMPARE the reproductive anatomy of the rhinoceros with that of the distantly related ***ZEBRA***. Although there are some similarities in their reproductive anatomy, such as a bicornuate uterus, there are also some notable differences. For example, zebras have external testes, whereas those of rhinos are internal.

Like most large mammals, rhinos are slow breeders. It is many years before they are mature enough to reproduce—more than six years in female white rhinos and even longer for male rhinos. After a long gestation period—well over a year for all species and up to 18 months in the white rhino—a single calf is born. The calf requires another 18 months of maternal care. The period between births may be up to five years. This slow birthrate makes rhino populations highly vulnerable if they are hunted.

The production and release of eggs in female rhinoceroses occur at regular intervals in response to changing hormone levels in the blood. When eggs are released from the ovaries, they pass down the fallopian tubes into the uterus, ready to be fertilized. At that time, females become sexually receptive. The time between periods of sexual receptivity is called the estrous cycle. In rhinoceroses, this ranges from 17 to 60 days, varying among individuals and species. The Sumatran rhino is an induced ovulator—it releases eggs ready for fertilization in response to sexual activity.

Bicornuate uterus

Apart from being large, the female reproductive organs are typical for a mammal. However, unlike that of humans, the rhinoceros uterus is bicornuate, having two well-developed tubes, or horns, that meet at a short central chamber leading to the entrance of the birth canal. In the white rhino, the uterus can be more than 3 feet (1 m) long. In addition, the rhino's ovaries are completely surrounded by the ends of the fallopian tubes, which lead from the ends of the uterine horns. In humans, the ovaries are open to the body cavity, and very occasionally fertilized eggs develop outside the uterus. In rhinos this

▼ Female and male white rhino

The female rhino's uterus is bicornuate (two-horn) and large—over 3 feet (1 m) long at full term. The ovaries are fully surrounded by the ends of the fallopian tubes. The male rhino's two testes are internal, and the penis curves backward between the hind legs.

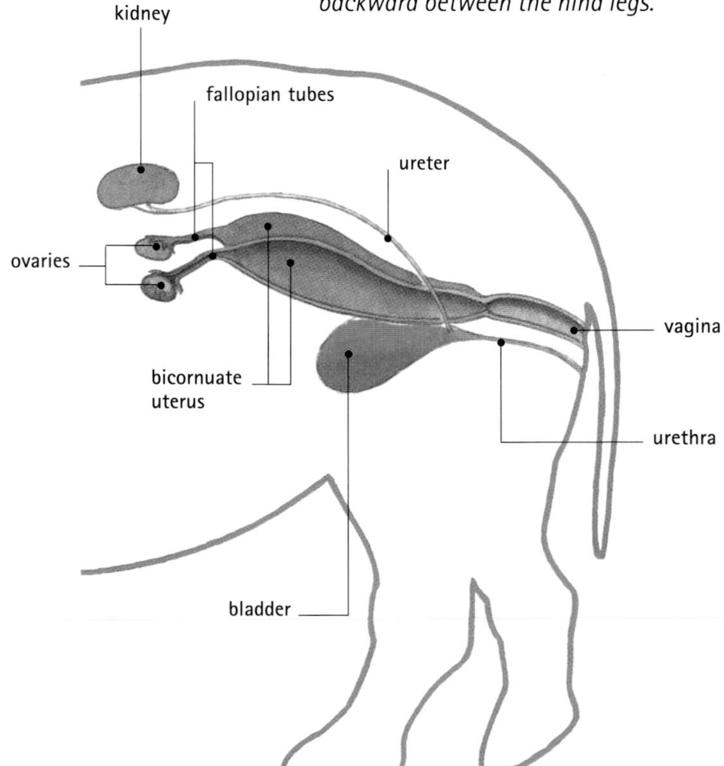

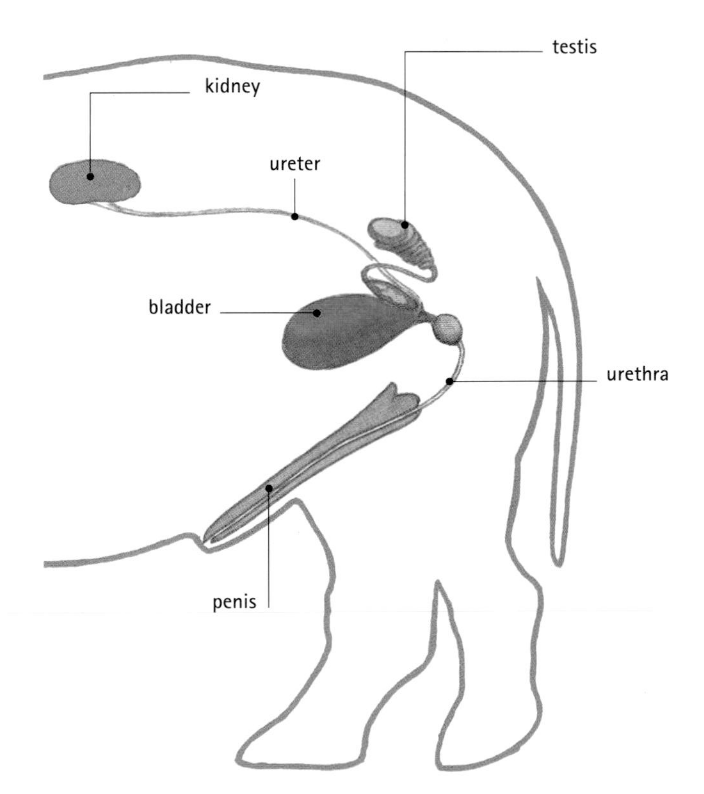

CLOSE-UP

Mating behavior

Breeding male rhinos of all species use urine to mark their presence in an area. Indian rhinoceroses also mark areas with scent produced by glands, called pedal glands, on the feet. However, not all species are territorial, and the ranges of breeding male Indian rhinos and black rhinos overlap. Indian rhinos are particularly aggressive, and dominant males will attack other males that they encounter. During courtship, female Indian rhinos will get into serious fights with persistent males. Although male and female white rhinoceroses sometimes chase each other and clash horns during courtship, there seems to be no such aggressive behavior between male and female black rhinos, which have a much gentler courtship.

▲ *A young white rhino stays with its mother for up to four years, or until she has another calf. A close bond between the mother and calf ensures that the youngster is well protected.*

cannot occur. The rhino embryo develops in one of the two uterine horns. At birth, a baby white rhinoceros can weigh up to 145 pounds (65 kg). Female rhinos have two milk-producing mammary glands located toward the groin area. A young rhino is suckled for one to two years depending on the species. In addition to the milk, black rhino calves begin nibbling vegetation within a few weeks; white rhinos' calves begin to nibble on grass at about three months old.

▼ **Suckling black rhinoceros calf**
A young black rhino calf suckles its mother's milk for about 18 months. It starts to eat grass and other leaves when it is a few weeks old.

Male reproductive system

The testes in male rhinos are always located inside the rhinoceros's body, near the kidneys. Several other mammals—including elephants, and their distant relatives the sea cows, dugongs, and manatees; and many other marine mammals—share this feaure. However, it is not characteristic of mammals in general. The reason why so many mammals have vulnerable external testes is that sperm stored in the testes—in a convoluted tube called the epididymis—survive better at cooler temperatures away from the warm body interior. Any male with more viable sperm would have a reproductive advantage over his competitors. Scientists are still uncertain why rhinos have internal testes.

A male rhinoceros's penis curves backward between its legs. This curious feature allows the rhino to direct urine for the purpose of scent-marking, which is a very important form of communication in rhinos.

ADRIAN SEYMOUR

FURTHER READING AND RESEARCH

Macdonald, David (ed.). 2006. *The Encyclopedia of Mammals.* Facts On File: New York.

Saguaro cactus

KINGDOM: Plantae ORDER: Caryophyllales
FAMILY: Cactaceae

Cacti form one of the most remarkable plant families. Their anatomy is suited to survival in extreme environments, from hot and arid deserts to frostbitten mountains. Cacti occur in a variety of forms, from huge treelike specimens to columns, mounds, and tiny plants that are almost completely buried in the soil. Most cacti have succulent stems and are armed with sharp spines. Several cacti are edible. Indeed, they have been part of the human diet for at least 9,000 years: mummified human feces in caves have been found to contain fragments of cacti. There is also a world trade in cacti as ornamental plants. The saguaro is the tallest cactus in the United States, the state flower of Arizona, and a symbol of the Sonoran Desert.

Anatomy and taxonomy

All life-forms are classified in groups that are part of larger groups. The classification is based mainly on shared anatomical features, which usually (but not always) indicate that the members of a group have the same ancestry. Thus the classification shows how the life-forms are related to each other. Scientists can also compare the anatomy of living organisms and fossil forms preserved in rocks of known age. The comparison can indicate how long the various groups have existed.

• **Plants** One of the key characteristics of most plants is that they are green. The color comes from a pigment called chlorophyll, which allows plants to create their own food from just air, water, and a few mineral nutrients (a process called photosynthesis). The other major groups of multicellular organisms (animals and fungi) cannot make their own food and all eventually depend on plants for their existence.

• **Flowering plants** Angiosperms, or flowering plants, are one of the two major groups of seed plants. Their ovules are inside an ovary, which ripens into a fruit containing the seeds. The other group, gymnosperms, includes pine trees and other conifers. Flowering plants are usually divided into monocotyledons (or monocots) and dicotyledons (dicots). Although no longer a strict taxonomic grouping, this division is nevertheless useful. Monocots generally have long, narrow leaves with parallel veins and flower parts grouped in threes. Monocots include palms, grasses (such as wheat and rice), and orchids. Agaves and aloes are monocots, even though, as succulents, they are superficially similar to cacti.

Dicots, or broad-leaved angiosperms, have leaves with netlike veining, and are very variable in shape. The seedlings usually have two cotyledons (seed leaves)—hence the group's name. The flowers usually have parts grouped in fours or fives. Most familiar plants, including roses and apple trees, are dicots. One group of dicots that looks similar to cacti but are unrelated is the euphorbias, many of which are spiny succulents.

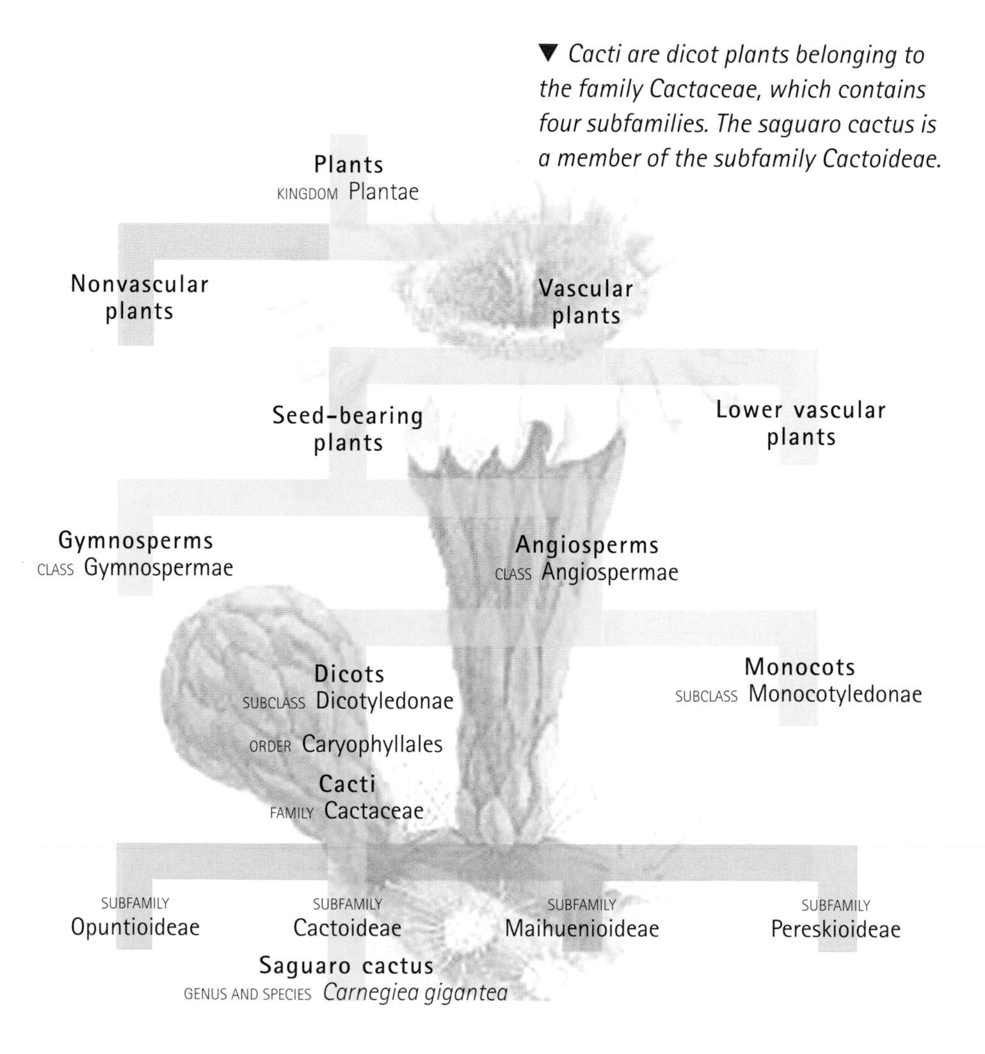

▼ *Cacti are dicot plants belonging to the family Cactaceae, which contains four subfamilies. The saguaro cactus is a member of the subfamily Cactoideae.*

• **Caryophyllales** Plants in the order Caryophyllales have distinctive pollen grains. Many members are succulents and produce pigments called betalains instead of anthocyanins. Betaleins are unique to Caryophyllales. Within the order are the spinach family (Chenopodiaceae); the carnation family (Caryophyllaceae); the stone plant family (Aizoaceae), which has many features in common with cacti; and the cactus family (Cactaceae).

• **Cactaceae** The cactus family has approximately 1,400 species in 97 genera. They probably diverged from other flowering plants around 70 million to 90 million years ago, in what is now the New World. Cacti occur naturally almost exclusively in the New World, from British Columbia to the southern tip of South America. They now also grow in many other countries where humans have introduced them. Although cacti typically inhabit hot and dry deserts, they live in a wide range of habitats including lush tropical rain forests and high mountains. This variety has led to a broad diversity of forms. Most characteristics of cacti are related to their need to conserve water. Most cacti are stem succulents: the leaves, where present, are simple (not divided into leaflets), succulent, very small, or short-lived. All cacti have areoles—flattened pads that represent axillary buds or short shoots and usually have spines.

There are four subfamilies of cacti: Pereskioideae, Maihuenioideae, Opuntioideae, and Cactoideae. Pereskias look nothing like cacti. They have broad leaves, and stems that are round, not fleshy. Close inspection, though, reveals the spine-bearing areoles that are characteristic of cacti. Maihuenioideae species are cushion- or mat-forming cacti from southern South America. They have persistent leaves, which are round in cross section.

Opuntioideae cacti have stems (cladodes) that are usually segmented by distinct joints. Leaves are produced but drop off after the growing season. These cacti have glochids (modified hairs with barbs) that are unique to the group. There are 15 genera including *Cylindropuntia*, *Opuntia* (prickly pear), and *Pterocactus*. The subfamily Cactoideae contains the "true" cacti, very variable plants with treelike, climbing, and spherical species. The leaves are absent or very small, and there are usually spines, but no glochids.

▲ *Large saguaro cacti in Saguaro Cactus National Park, Arizona. Some individuals are more than 100 years old.*

FEATURED SYSTEMS

EXTERNAL ANATOMY Cacti have leaves modified to form spines, which grow from pads called areoles. The swollen stem stores water. The saguaro cactus is a very large columnar branching species, which commonly grows to 40 feet (12 m) tall. *See pages 1050–1054.*

INTERNAL ANATOMY Cactus stems are packed with water-storing parenchyma cells. With no leaves, the outer cells of the stem are photosynthetic. A thick waxy cuticle reduces the rate at which cacti lose water in arid environments. *See pages 1055–1057.*

REPRODUCTIVE SYSTEM Cacti have brightly colored flowers with many petals. The fruits are often juicy, and the seeds within are dispersed by animals. *See pages 1058–1059.*

• **Saguaro cactus** The saguaro is the largest cactus in the United States and commonly reaches 40 feet (12 m) tall. One was measured at 78 feet (23.8 m). The saguaro is a columnar branching cactus, with cylindrical stems 12 to 30 inches (30–75 cm) in diameter. The stems are "pleated," and the ridges (ribs) have areoles, each with 15 to 30 straight spines. White flowers grow near the top of the branches and are followed by red, fleshy fruits. Native Americans value the saguaro because it fruits at a time of year when little other food is available.

External anatomy

CONNECTIONS

COMPARE the spines of a saguaro cactus with the scalelike leaves of a ***SEQUOIA***. Both structures are true leaves, but they are modified in different ways. In sequoia, the leaves are reduced to scales, giving a smaller surface area than normal leaves. Scales remain, however, the plant's primary site of photosynthesis. In cacti, the leaves have been reduced and toughened to form spines and have no chlorophyll, so they cannot photosynthesize.

Cacti are superbly suited to life in hot, arid environments. They form one of several plant families that have swollen, succulent stems, absent or reduced leaves, and clusters of spines. Cacti come in a huge variety of shapes and sizes. *Blossfeldia liliputana* is one of the tiniest species at about 0.4 inch (0.9 cm) in diameter, whereas *Cereus lamprospermus* grows to more than 60 feet (19 m) high. Some climbing species may reach even higher.

Some cacti are treelike, with a trunk and several branching stems. A distinctive form is the candelabra: the trunk divides into several vertical branches similar to a branching candlestick. Shrubby cacti, with several stems arising at or near ground level, include *Stenocereus thurberi* and members of the mat-forming genus *Maihuenia*.

The saguaro is one of many columnar cacti. These plants may be branched, like saguaro, or unbranched. Globose, or globular, cacti are spherical in shape, sometimes with a flattened top (for example, *Echinocactus platyacanthus* and *Gymnocladium hossei*). Cacti may also be solitary or clustering (caespitose), with many stems coming from a common base. The clusters can be tight, as in *Copiapoa conglomerata*; or loose, as in *Echinocereus cinerascens*.

Some cacti clamber over the soil surface (for example, *Stenocereus eruca*), and others are climbers (for example, *Hylocereus undutus*). The latter cling to trees or rocks with aerial roots or by twining. Some cacti do not even grow in soil: they can survive clinging to bare rock (lithophytes) or to other plants (epiphytes). About 10 percent of all cacti are epiphytes.

▶ *The golden barrel cactus grows in Mexico. It grows taller and more barrel-shaped with age, and large specimens may be 3 feet (0.9 m) tall.*

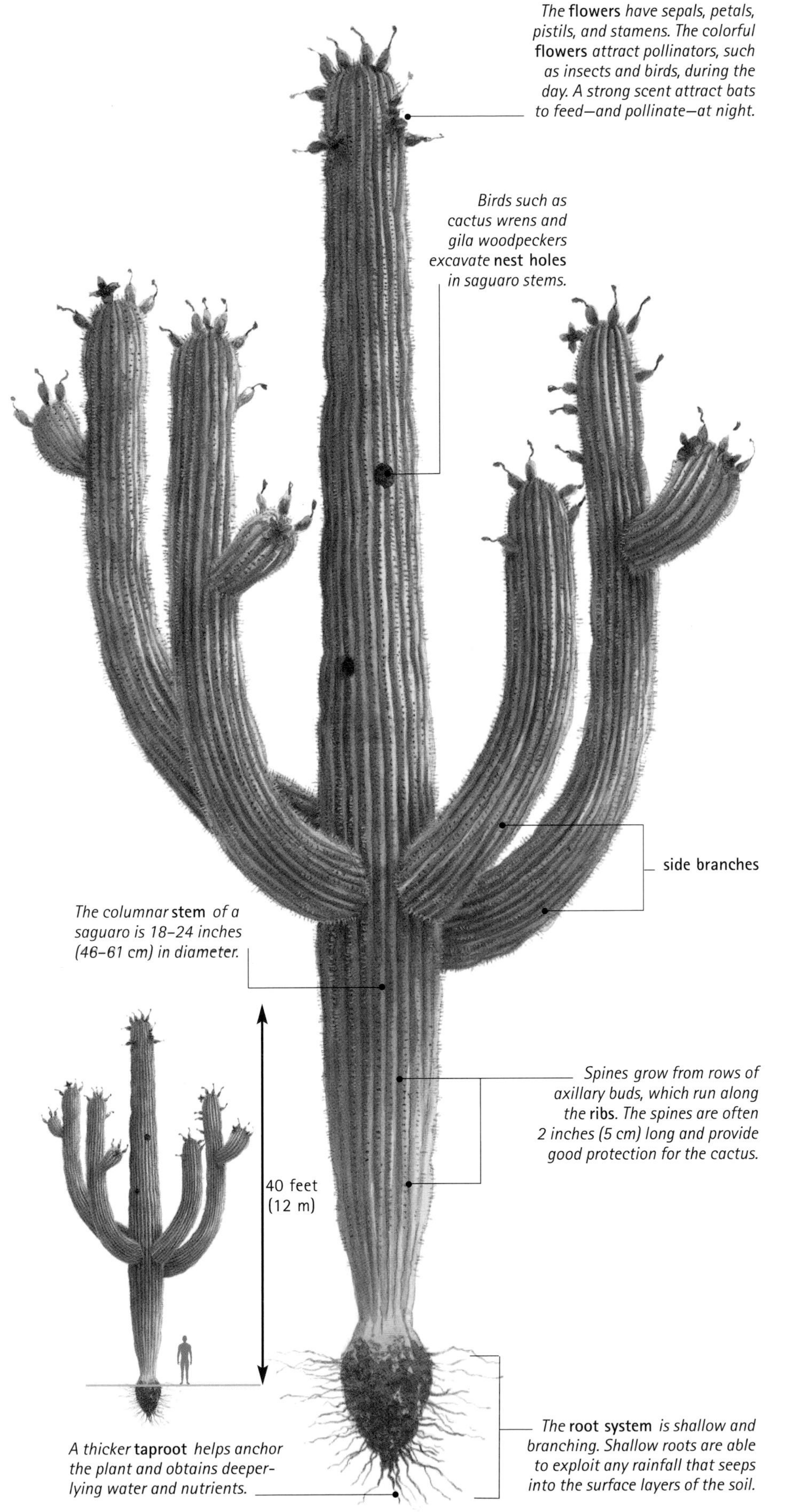

◀ **Saguaro cactus**

The saguaro cactus is the state flower of Arizona. It lives in the Sonoran Desert of southeastern California, southern Arizona, and parts of northwestern Mexico. Some individuals grow more than 40 feet (12 m) tall.

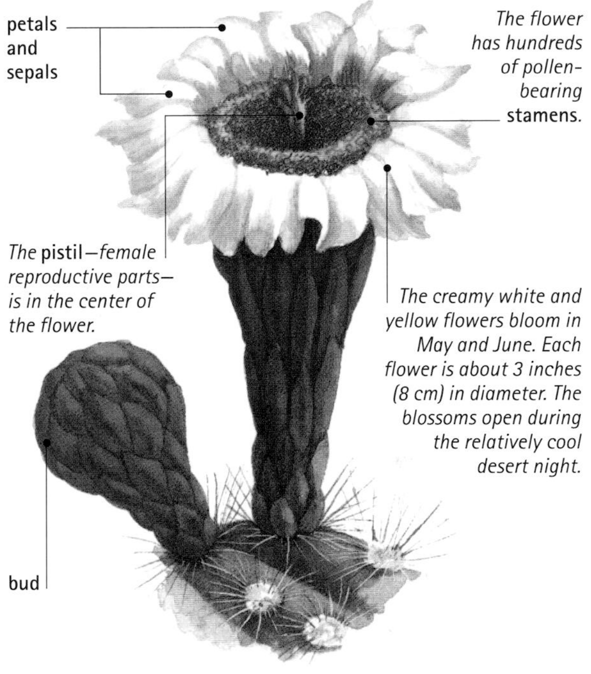

▲ FLOWER

Saguaro cactus flower

The flowers of a saguaro cactus have a structure like that of other flowering plants. Saguaros can be fertilized only by cross-pollination: pollen brought from other saguaros by insects, birds, or bats that are attracted to the flowers' color and odor.

External structures

Stems form the bulk of the aerial parts of cacti. In most cacti, the stems are the main site of photosynthesis (in most other plants the leaves serve this role). In most cacti, the stem also stores water, and 90 percent of a cactus's weight may be water; the plant can survive even with a water content as low as 20 percent.

Cactus stems are usually cylindrical, but they may also be flattened, as in many *Opuntia* species. When such flattened stems are distinctly jointed, as in the prickly pear cactus, they are called cladodes. The cladodes are sometimes mistaken for leaves, but—unlike true leaves—they bear areoles (modified buds) and can grow new branches. The ribs of

EVOLUTION

Convergent evolution

Unrelated plants often cope with similar environmental pressures by adopting similar growth forms. This process is called convergent evolution. The columnar shape and spiny, leafless form of some African euphorbias are similar to the shape of many American cacti, although the two groups of plants are not closely related.

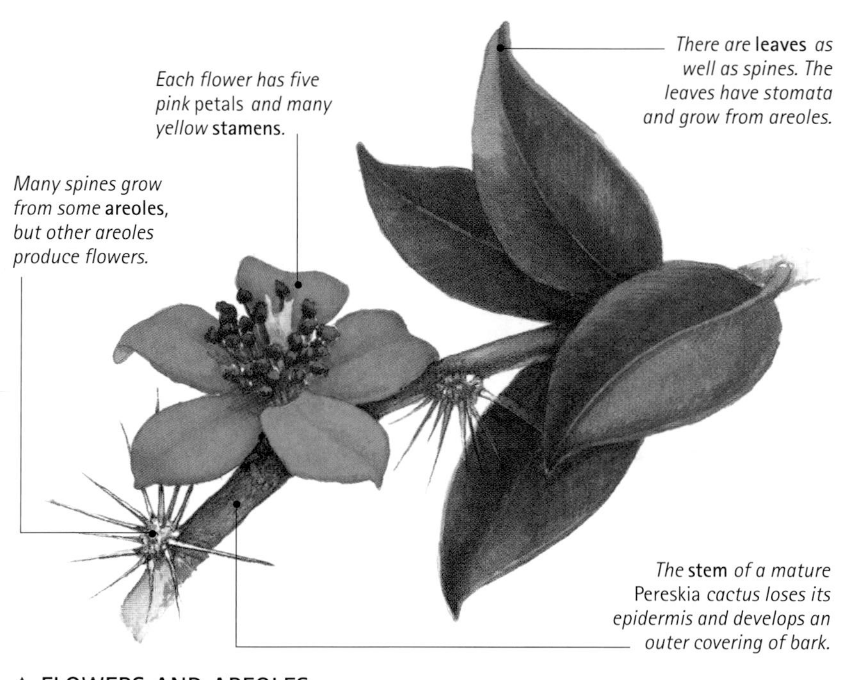

▲ FLOWERS AND AREOLES
Pereskia diaz
The stems of mature Pereskia *cacti are unusual in having woody bark, with few stomata. However, the leaves do have stomata.*

columnar and some globose cacti have concertina-like edges that can expand rapidly after rain (over hours or days) without bursting or having to grow new cells.

Areoles develop on tubercles, protuberances that may be arranged around the stem in crisscrossing spirals or merge into vertical ribs, as in saguaro. *Mammillaria longimamma* has long, nipplelike tubercles, whereas those of some *Ariocarpus* species are triangular and leaflike.

Most cacti lack typical flattened, green leaves. *Pereskia* species, and also some members of the subfamily Opuntioideae, have fleshy, persistent green leaves or ephemeral leaves that are dropped after the growing season. True cacti in the subfamily Cactoideae have no visible leaves, although the plants do produce microscopic vestigial leaves.

Areoles, spines, and hairs

Areoles are unique to cacti. An areole is a highly modified lateral bud or dwarf shoot that produces spines and sometimes hairs or flowers. Spines are modified leaves—compare them with thorns, which are outgrowths of branches. Some cacti have spines only when they are seedlings (for example, *Peyote* and *Lophophora* species), but most are spiny throughout their life. Spines vary in number, size, shape, and color, so they are often used in cactus identification. They can be thin and hairlike (as in *Cephalocereus*), flattened and papery (for example, *Maihueniopsis glomerata*), or hooked (*Ferocactus*, for example).

In saguaro cactus, the areoles are distributed at intervals of about 1 inch (2.5 cm) along the ridges of the ribs. Each areole bears a cluster of about 30 spines, the longest of which are up to about 2 inches (5 cm). On young plants the spines are stout and sharp. Once stems grow

▼ FRUIT
Saguaro cactus
When ripe, the saguaro fruit is a red and fleshy berry with small black seeds. The fruits are eaten by birds and mammals, which disperse the seeds.

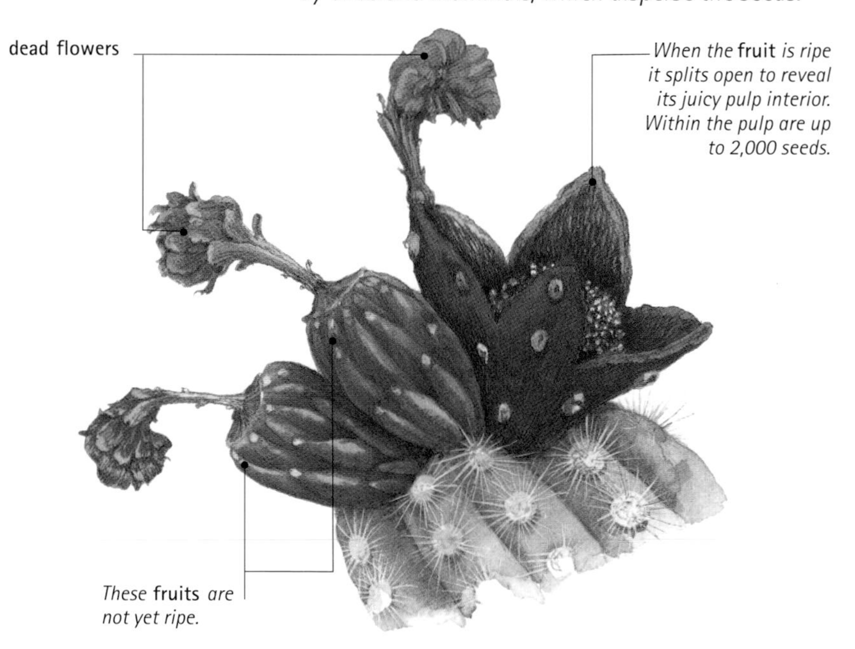

CLOSE-UP

The cephalium

Some types of cacti have a specialized flower-bearing area, which forms only when the cactus is mature and ready to flower. This area, called the cephalium, has an anatomy different from the rest of the stem. The cephalium has very short internodes (the distance between areoles) and is usually densely hairy or woolly. In most species (for example, *Melocactus*), the cephalium forms a cap at the top of the stem; but in others (for example, *Epostoa*), it forms a band down the side, taking in several ribs.

▼ Melocactus
This low-growing species has a large cephalium.

taller than about 8 feet (2.4 m) they produce bristly spines. Hairs may give the areole a woolly appearance, as in *Mammillaria hahniana*. Opuntias produce modified hairs called glochids, which are stiff and barbed and detach easily, even from the lightest brush of a human hand. Because they are barbed, glochids lodge easily in skin and are very painful.

Spines serve many purposes. Most important, they protect the cactus from grazing animals. This protection is essential because in arid environments vegetation is very limited, and anything green becomes a target for hungry and thirsty animals. Cacti spines may also act as camouflage, as in *Sclerocactus papyracanthus,* whose long, papery spines disguise the cactus as dead grass. Spines also shade the stem surface by reflecting light, thus reducing water loss by evaporation. They also hold a layer of still air (a boundary layer) around the plant, minimizing the drying effects of the wind. For cacti that live in areas such as

IN FOCUS

Surface area and desert survival

Cacti have capitalized on some basic laws of math to ensure their survival in extreme conditions. The ratio of a plant's surface area to its volume affects its interactions with the environment. A plant gains and loses heat through its surface, and it also loses water through stomata, or air pores. With no true leaves, cacti exchange air through stomata on the stem. The stomata are often located in sunken pits. This arrangement minimizes air movement and, therefore, water loss.

The surface (or at least the outer fraction of an inch) is also the only photosynthetic area. The body of the cactus (its volume) is used to store water and also acts as a heat regulator. Heat absorbed during the day is stored in the aqueous tissue mass and slowly released back to the air at night, so that temperature fluctuations in the plant tissues are much less extreme than in the air around the cactus. For most plants, photosynthesis to make food is a priority. Therefore they produce lots of leaves with a large surface area and low volume. Cacti, however, live in places where sunlight is usually plentiful, but water is in short or irregular supply, and temperature control is critical. Unlike most plants, cacti therefore require a very low surface area to volume ratio (SA:vol). The shape that provides the lowest SA:vol is a sphere, and many cacti have this form. A thick column also has a low SA:vol.

the Atacama Desert, Chile, where there is frequent fog, the spines provide a large surface area on which dew can condense and drip onto the soil at the plant's base.

For cacti with hairs as well as spines, such as montane cacti, the hairs serve as insulation, trapping warm air when night temperatures can fall below freezing. White hairs also reflect sunlight, giving protection from damaging ultraviolet rays as well as providing cooling shade for the underlying tissues. Without spines or hairs, the surface of a saguaro cactus can reach 122°F (50°C) in full sunshine. Together, shading by spines and hairs can reduce this temperature by 57°F (14°C).

The root system

Roots take up water and minerals, anchor the plant in soil, and store water and carbohydrates. Saguaro and other large columnar cacti have one main taproot that penetrates vertically to anchor the plant, as well as absorb water and nutrients. The taproot can extend to more than 2 feet (60 cm) deep. Some smaller species (for example, *Copiapoa*) have a large taproot, shaped like a carrot, that stores water.

The remainder of the root system is extensive and shallow, as is the case for most succulent plants. The lateral roots of a saguaro cactus are rarely more than 4 inches (10 cm) deep and radiate horizontally about as far from the plant as the plant is tall. This extensive network of shallow roots quickly captures any rain that falls. Some slender climbing or scrambling cacti do not use their stems for storage but rely on swollen roots that act as underground storage organs, like tubers (true tubers are underground stems).

▶ *This African euphorbia is not a cactus, but it shares important features with the saguaro cactus. Euphorbias and cacti grow in arid environments, and both have evolved water-saving features: neither group has leaves, and in both groups the stems contain water-storage tissues. This is an example of convergent evolution between the two groups of plants.*

Internal anatomy

The internal anatomy of cacti is based on the same structures that are found in other dicotyledonous plants, namely an epidermis, stomata, a cortex, vascular bundles, and pith. However, this basic structure is modified in various ways that enable cacti to cope with the extreme environments in which they live.

The outer surface of a cactus stem is usually covered by a thick waxy cuticle. Species in the genera *Ariocarpus* and *Copiapoa* have a waxy layer that may be 0.0005 inch (20 µm) or more thick. In most deciduous trees, the waxy layer is only 0.000025 to 0.00015 inch (1–6 µm) thick. The epidermis is the outermost layer of cells, and these are tightly packed. Below the epidermis is the hypodermis, a layer that is usually several cells thick. The cells are thick-walled and tough and provide mechanical protection for the stem.

Below the hypodermis, the cortex consists of a pithy core that makes up the bulk of the cactus stem. The outer layer of the core, called the chlorenchyma, is a fraction of an inch thick and green. Its color comes from chloroplasts—the photosynthetic organelles that pack its cells. Chloroplasts contain the green pigment chlorophyll. This pigment is responsible for capturing energy from sunlight to make sugars—the process called photosynthesis.

Water-storage cells

In the center of the stem, and making up most of its bulk, are parenchyma cells. These are large, loosely packed cells with no chloroplasts.

CONNECTIONS

COMPARE the stem of a saguaro cactus with the trunk of an ***APPLE TREE***. In an apple tree, the bulk of the trunk is wood. In a saguaro cactus only the core is woody, and the rest consists mainly of water-storing parenchyma. The bark of an apple tree is rigid, but the ribs of a saguaro cactus allow the stem to expand outward to store water, and to contract during periods of drought.

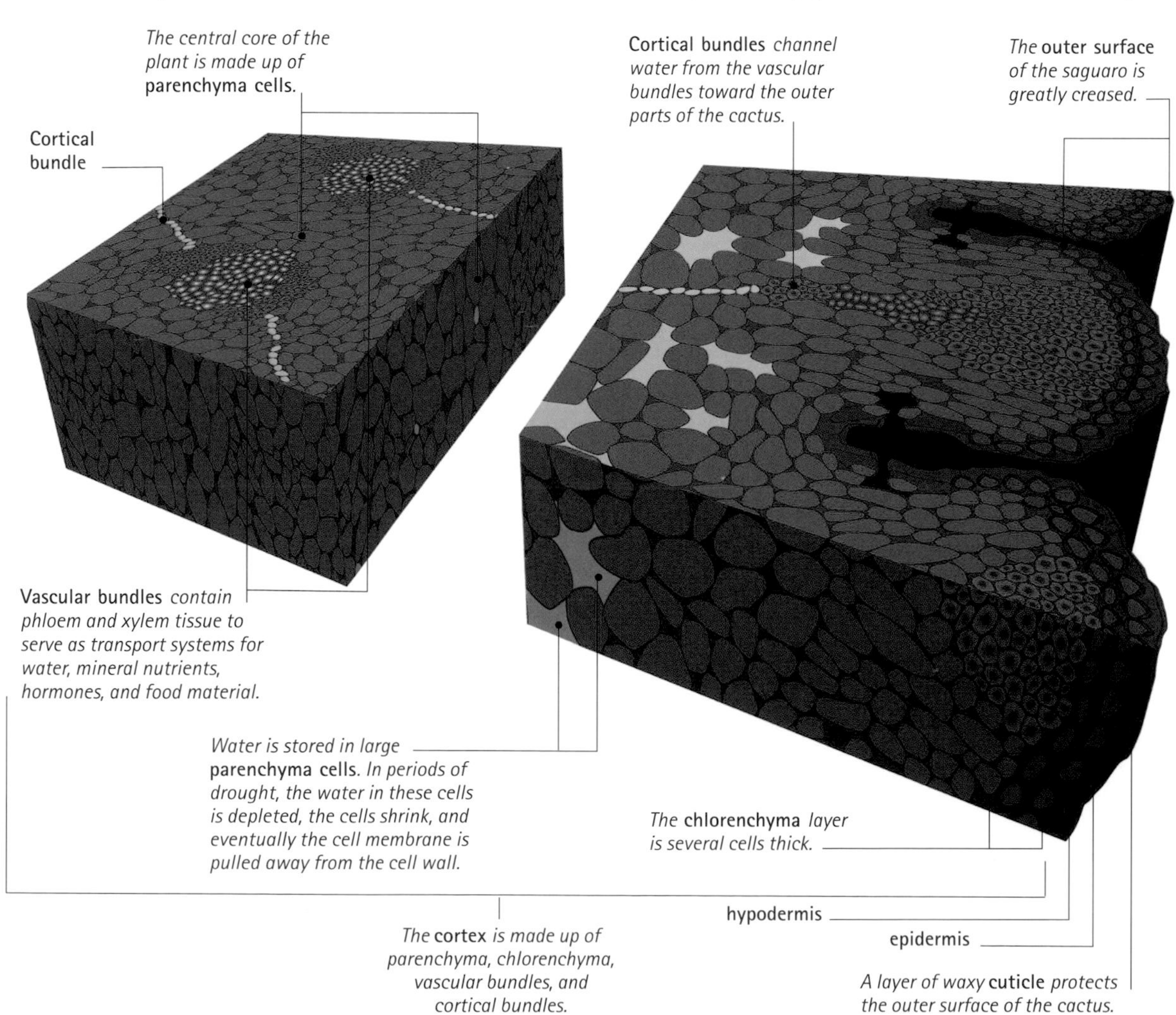

◀ STEM CROSS SECTION
Saguaro
The outermost layer of the stem, the epidermis, is protected by a waxy cuticle. Below the epidermis is the hypodermis, and within this is the cortex, which makes up the bulk of the saguaro's volume. The cortex is made up of chlorenchyma and water-storing parenchyma. In addition, there are vascular bundles and cortical bundles in the cortex. The latter provide a lateral water transport system.

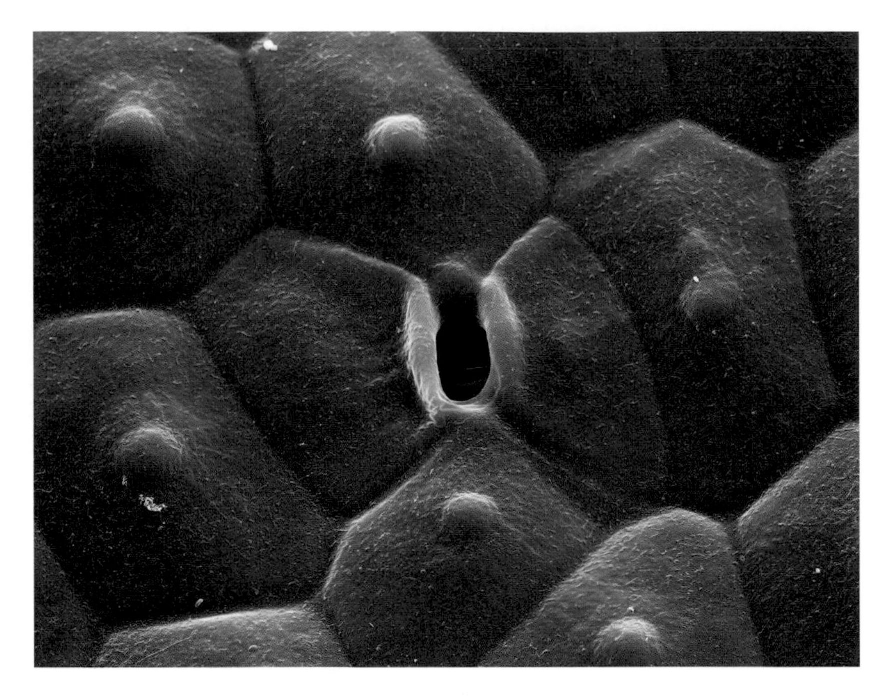

▲ *A stoma is clearly visible as a hole in the surface of this cactus. The stoma is opened or closed by two bordering guard cells. Magnification x 400.*

Water is stored in the parenchyma cells. Vascular bundles form a ring within the stem and consist of clusters of xylem and phloem. Xylem vessels conduct water and mineral nutrients from the roots, and phloem conducts the sugary products of photosynthesis throughout the plant. In the saguaro and other large species of cacti, the profuse growth of thick-walled xylem vessels forms woody tissue that supports the plant. The large regions of parenchyma cells, which are necessary to meet the cacti's water-storage demands, result in wedges of wood separated by large areas of softer tissue. The resulting wood cores, with a star-shape cross section, are visible in dead saguaros when the parenchyma has decayed.

Coping with drought

Most plants open their stomata during the day to take in the carbon dioxide needed for photosynthesis. However, if a plant's stomata are open in hot sunshine, the plant loses large amounts of water. Cacti need to conserve water, and they have a specialized chemical pathway that enables them to take in carbon dioxide during the night, when the air is cooler, and store it for use during the day.

Carbon dioxide is converted into organic acids, which are stored in the vacuoles in chlorenchyma cells. Vacuoles are membrane-bound sacs inside cells that contain mostly water with a few dissolved compounds (solutes). Once the sun rises, the cactus closes its

▼ CEPHALIUM SECTION
Melocactus
The cephalium contains meristem tissue, which gives rise to flowers.

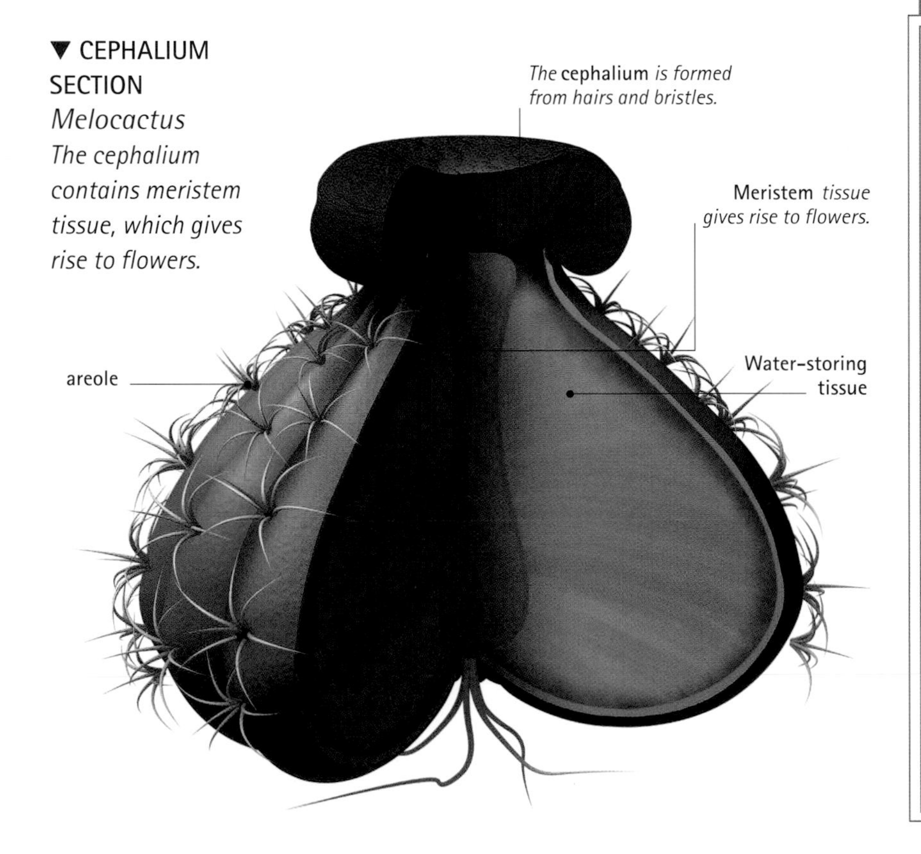

CLOSE-UP

Cells and drought

Cacti can withstand long periods without rain by storing water in the parenchyma cells of their stems. During a long drought, water from the central parenchyma cells is moved to the chlorenchyma cells, where it is used in photosynthesis. As water is slowly used up, the parenchyma cells shrink. Their relatively rigid cell walls start to collapse, giving the cells an irregular shape, and air gaps develop between the cells. As the drought continues, the cell's contents shrink further and air gaps appear within the cells, pulling the cell membrane away from the cell wall. As the water content of the cell drops, the concentration of solutes (dissolved molecules such as sugar) increases. For cells of most plants, a water loss of 30 percent is usually lethal. The parenchyma cells of cacti, however, can tolerate losses of 70 to 95 percent before high solute concentrations cause lethal damage.

IN FOCUS

From birds to "boots"

Saguaros make excellent nesting places for many birds. Gila woodpeckers and gilded flickers can easily excavate nesting holes in the fleshy stems. The cactus quickly lines the excavation wounds with callus—rapidly grown, undifferentiated cells that dry out to form a corky protective layer. When the saguaro eventually dies, this callus tissue decomposes more slowly than the rest of the cactus, so the remains of the nest holes can be found on the ground among the debris of dead plants. Because of their shape, they are called "saguaro boots." Native Americans sometimes use them as containers.

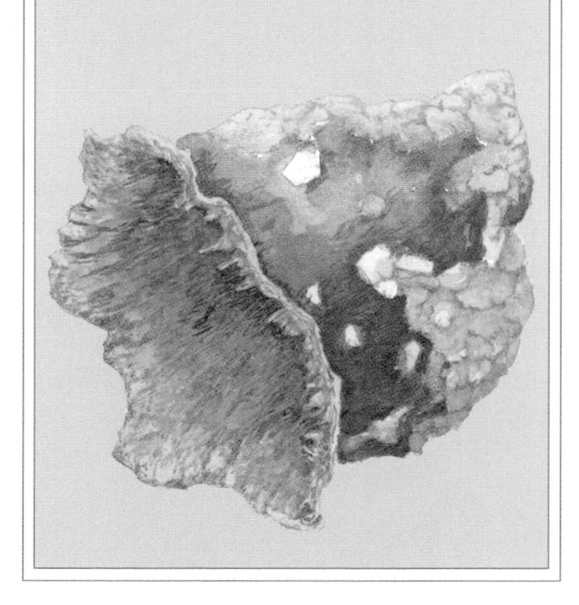

▼ Saguaro boot

Desert animals such as snakes, scorpions, and spiders sometimes live in a saguaro boot. The boots can collect water in the rainy season and may act as tiny sources of water for a short time.

stomata to limit gas exchange as well as water loss, and the stored acids are broken down so the carbon can be used to make sugars in photosynthesis. This chemical pathway was first discovered in a plant in the genus *Crassula*, so it is called crassulacean acid metabolism (CAM). Scientists believe that more than 20,000 plant species use CAM. In addition to desert cacti, these include many orchids.

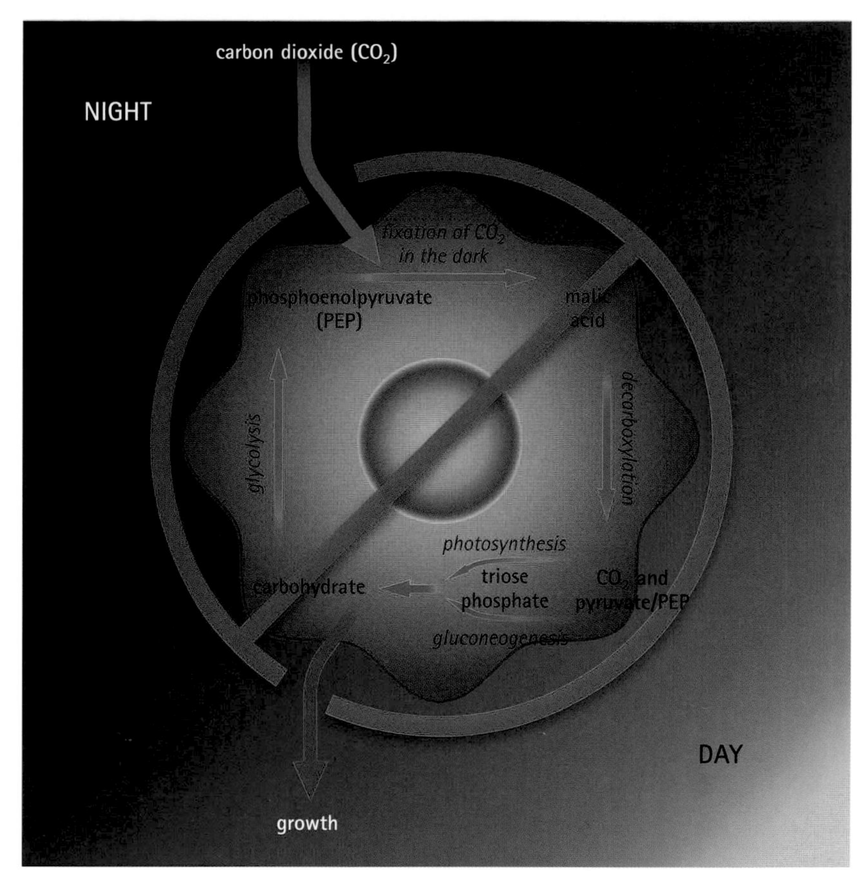

▲ CRASSULACEAN ACID METABOLISM

Many plants living in arid conditions, including cacti, use CAM, in which stomata open only at night. The carbon dioxide entering the stomata is stored as an acid before being used in light-dependent photosynthesis the following day.

▶ *This gilded flicker has constructed a nest hole several feet up in the stem of a saguaro cactus. There, the bird's eggs and chicks are relatively safe from predators. The cactus "heals" the inside of the hole with a hard layer of callus.*

Reproductive system

CONNECTIONS

COMPARE the flowers of a cactus with those of an ***APPLE TREE***. In an apple blossom, the petals and sepals can be easily distinguished: sepals are small and green; petals are large, delicate, and colored. In cacti, the perianth (sepals and petals) are not easily distinguished.

Cacti depend predominantly on sexual reproduction for creating new plants, although asexual reproduction plays a minor role in some species. In many *Opuntia* cacti, the cladodes can break off and root where they land, establishing new plants. However, flowers and the resulting fruits and seeds are the prime means by which cacti reproduce.

Flowers and fruits

Cactus flowers tend to be large, solitary, and colorful, often in shades of yellow, magenta, red, or white. They arise from areoles, usually near the apex of the plant but sometimes lower down. Saguaros do not flower until they are large and at least 40 years old. White flowers form mostly near the stem tips in May and June. The sturdy flowers are about 3 inches (8 cm) in diameter.

All flowers have the same basic components. On the outer edge, closest to the stem, is a whorl of sepals, which protect the developing bud. Sepals are usually robust and green. Together they form the calyx. Within the calyx is a whorl called the corolla, made up of petals. Petals are usually flimsier than sepals and are brightly colored. Together, the calyx and corolla are called the perianth. Most dicot flowers (for example, apple and Venus flytrap flowers) have petals and sepals grouped in fours or fives, whereas monocot flowers (for example, grasses, irises, and orchids) have flower parts grouped in threes. Cactus flowers are unusual:

IN FOCUS

Seed dispersal

Saguaro seeds, like those of many cacti, are dispersed primarily by fruit-eating birds, such as white-winged doves, gila woodpeckers, and house finches. The birds eat the pulp, and the seeds pass through their digestive system. The fruits are also eaten by long-nosed bats. In some of the larger fruited cacti, the fruits fall to the ground, where they emit a scent. The scent attracts feeding mammals or reptiles. Cacti with spiny fruits—for example, *Cylindropuntia* and *Opuntia*—can attach to a mammal's fur or a bird's feathers. In a few species, such as *Blossfeldia*, the seeds are dispersed by ants.

▶ *Lesser long-nosed bats pollinate saguaros at night (right). The bats have a long tongue, which enables them to feed on nectar in the saguaro flower. As they feed, pollen attaches to the bats' fur and is carried to the next cactus. Honeybees pollinate the cacti during the daytime (far right).*

▲ *The claret cup cactus has bright red flowers that attract hummingbirds. To reach the nectar the birds' whole head enters the flower cup and becomes dusted with pollen.*

although they are dicotyledons, there is no fixed number of floral parts. In addition, there is no clear distinction between sepals and petals; instead, the flower parts show a gradual transition in form.

Inside the perianth are the structures that make up the male and female portions of the flower. The male parts are the stamens. A stamen consists of a long filament with an anther on top, in which the pollen develops. The flowers of most cacti have numerous stamens. At the center of the flower is the female structure, called the pistil. At the base of the carpel is the ovary, which contains the ovules. These will develop into seeds when fertilized. In cacti, the ovary is sunk into the stem tissue (called the pericarpel). Above the ovary is a long stalk called the style, and at the top of the style is the stigma, the receptive surface on which pollen must land for eventual fertilizaton of the ovules into seeds.

The saguaro cactus fruits are berries. They mature in June and July and are about 3 inches (8 cm) long and bright red when ripe. The berries split open to reveal up to 2,000 small black seeds in red juicy pulp. All cacti produce numerous seeds, which are usually embedded in a fleshy pulp. Cactus seeds vary in size, shape, color, and seed-coat pattern, and they serve as useful guides for identifying the different species of cacti.

ERICA BOWER

IN FOCUS

Pollination in cacti

Saguaro flowers share many characteristics with other flowers that are pollinated by bats. The robust white flowers open at night and emit an aroma like that of overripe melon, a fragrance that is attractive to bats. However, the flowers remain open for most of the following day, and are also pollinated by insects, mainly bees. Other cacti with night-opening flowers—for example, senita—are pollinated by moths. *Rebutia* species are adapted for butterfly pollination, with a "landing-pad" lip and long tubular flowers where the nectar can be reached only by insects with a long proboscis. Some cacti have evolved to be pollinated by hummingbirds. These cacti usually have bright red flowers with no scent, copious nectar, and long floral tubes.

FURTHER READING AND RESEARCH

Anderson, E. F. 2001. *The Cactus Family*. Timber Press: Portland, OR.

Bell, Adrian D. 1991. *Plant Form: An Illustrated Guide to Flowering Plant Morphology*. OUP: Oxford, UK.

Heywood, V. H. 2007. *Flowering Plant Families of the World*. Firefly: Toronto.

Noble, P. S. 1994. *Remarkable Agaves and Cacti*. OUP: Oxford, UK.

The Arizona-Sonora Desert Museum: www.desertmuseum.org/books/cacti.html

Sailfish

CLASS: Osteichthyes ORDER: Perciformes
FAMILY: Istiophoridae GENUS: *Istiophorus*

The sailfish, the world's fastest fish, is a high-speed predator that eats fish and squid. It roams the surface waters of the tropical and warm temperate Atlantic, Indian, and Pacific oceans. The sailfish is a kind of billfish: it is armed with a spearlike bill. The sailfish probably uses sideways swipes of its bill to stun or injure prey fish before eating them.

Anatomy and taxonomy

Scientists categorize all organisms into taxonomic groups based partly on anatomical features. The sailfish is one of 11 species of billfish belonging to the families Istiophoridae and Xiphiidae. These are fast-swimming fish with an upper jaw extended into a spearlike point. These families, in turn, belong to the suborder Scombroidei (tunalike fish) within the order Perciformes (perchlike fish), which is by far the largest order of ray-finned fish.

- **Animals** Sailfish, like other animals, are multicellular and get food by eating other organisms. Animals differ from other multicellular life-forms in their ability to move from one place to another (in most cases, using muscles). They are generally able to react rapidly to touch, light, and other stimuli.

- **Chordates** At some time in its life cycle, a chordate has a stiff, dorsal (back) supporting rod called a notochord that runs all or most of the length of the body.

- **Vertebrates** In living vertebrates, the notochord develops into a backbone (spine or vertebral column) made up of units called vertebrae. The vertebrate muscular system that moves the body consists primarily of muscles in a mirror-image arrangement on either side of the backbone or the notochord (this arrangement is called bilateral symmetry about the skeletal axis).

- **Gnathostomes** These are jawed fish, as opposed to agnathans, such as hagfish and lampreys, which lack proper jaws. Gnathostomes have gills (the breathing apparatus) that open to the outside through slits, and fins that include those arranged in pairs, such as the pectoral (shoulder) fins.

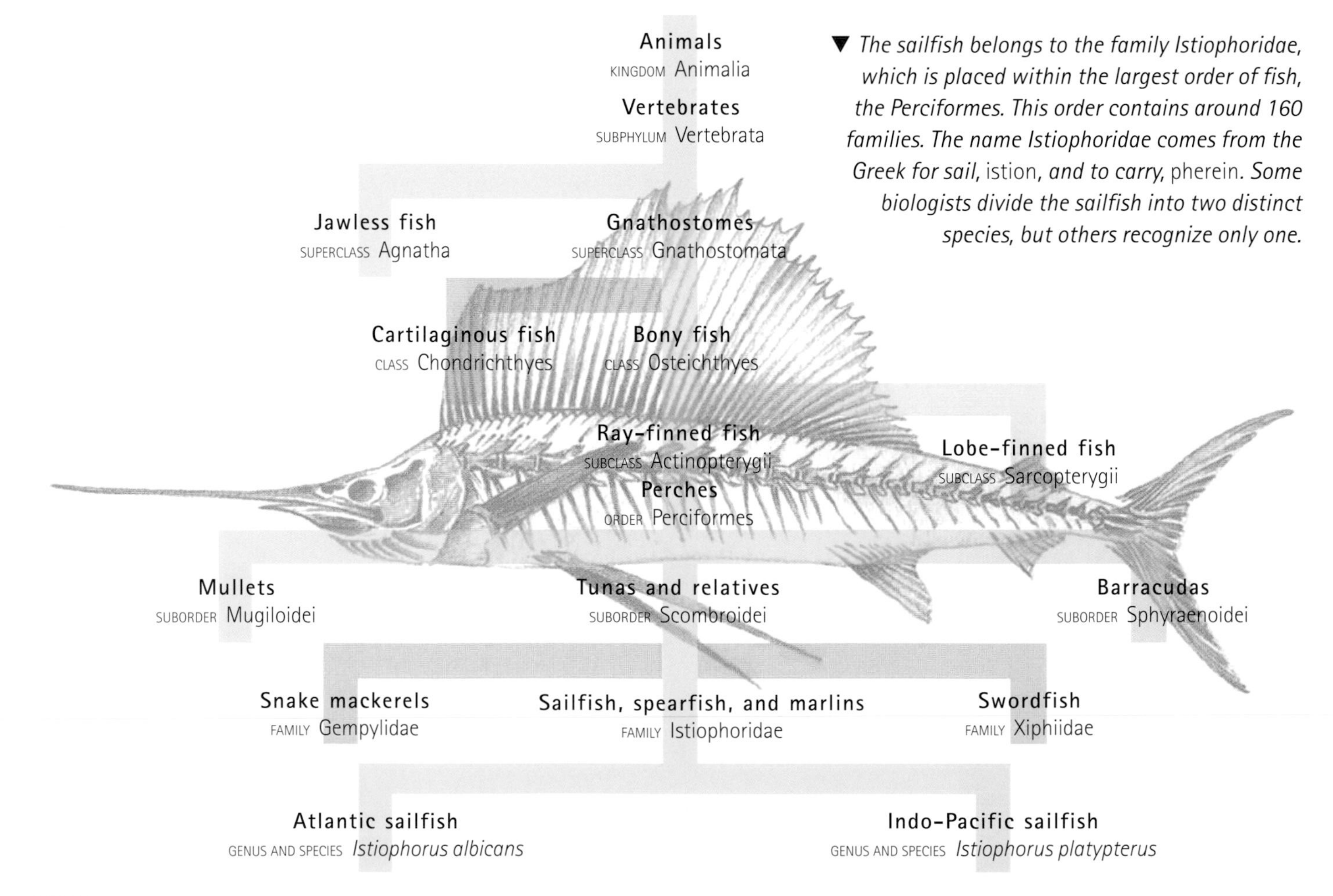

▼ *The sailfish belongs to the family Istiophoridae, which is placed within the largest order of fish, the Perciformes. This order contains around 160 families. The name Istiophoridae comes from the Greek for sail,* istion, *and to carry,* pherein. *Some biologists divide the sailfish into two distinct species, but others recognize only one.*

▶ *A sailfish cruises in the upper ocean waters looking for prey. The sailfish has its sail-like dorsal fin only partially erect.*

• **Bony fish** Sailfish belong to the class Osteichthyes (bony fish), which is the major group that includes more than 95 percent of all fish. Bony fish, as their name implies, have a skeleton of bone, in contrast to members of the class Chondrichthyes (cartilaginous fish), such as sharks, rays, and chimaeras, which have a skeleton made of cartilage.

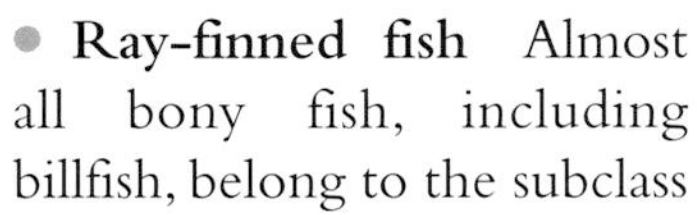

• **Ray-finned fish** Almost all bony fish, including billfish, belong to the subclass Actinopterygii (ray-finned fish). The major feature that distinguishes them from species of the subclass Sarcopterygii (fleshy-finned fish) is the presence of long bony rays that provide support for thin, flexible fins.

• **Perches and their relatives** The order Perciformes is a hugely abundant group of ray-finned fish. It contains about 10,000 species in some 160 families and is by far the biggest and most diverse order of vertebrates.

• **Tunas and their relatives** Also called the scombroids, these are members of the suborder Scombroidei. They are fast-swimming oceanic fish that have an upper jaw in which the bones are fused to form a beak, rather than being extensible (able to be pushed outward) as in other perchlike fish. Apart from tunas and mackerels, other scombroids include barracudas, snake mackerels, cutlass fish, and billfish.

• **Billfish** Billfish are renowned for their speed and strength. These large tunalike fish have an upper jaw extended into a long bill. In the sailfish, spearfish, and marlins the bill is rounded in cross section, and there are two ridges, called horizontal keels, on each side of the base of the tail fin. In the swordfish the bill is slightly flattened horizontally and there is one horizontal keel at either side of the tail fin. The sailfish differs most obviously from other billfish in its extended sail-like first dorsal fin, from which it gains its name. Some authorities recognize two species of sailfish: Atlantic sailfish and Indo-Pacific sailfish.

FEATURED SYSTEMS

EXTERNAL ANATOMY The highly streamlined body and narrow fins are characteristic of fish that cruise at high speed. *See pages 1062–1064.*

SKELETAL SYSTEM Long slender bones provide support for the sail-like dorsal fin. *See page 1065.*

MUSCULAR SYSTEM A tendon–pulley system generates powerful side-to-side strokes of the tailfin, which thrusts the fish forward. *See pages 1066–1067.*

NERVOUS SYSTEM Vision is the sailfish's most important sense. A modified eye muscle generates heat that keeps the eyes and brain warm for rapid operation even in cool water. *See pages 1068–1069.*

CIRCULATORY AND RESPIRATORY SYSTEMS Like other billfish, the sailfish has a large heart, and blood with a high oxygen-carrying capacity. This is needed for delivering oxygen to the contracting red muscle that maintains the fish's high cruising speed. *See pages 1070–1071.*

DIGESTIVE AND EXCRETORY SYSTEMS Billfish consume their prey whole and require an efficient digestive system to break down large food items. *See page 1072.*

REPRODUCTIVE SYSTEM Females produce millions of eggs that are fertilized externally. Most of the eggs and young larvae are eaten by predators, with only one, or a few, on average growing to adulthood. *See page 1073.*

External anatomy

Like other billfish, the sailfish has a long snout, or bill, formed from the elongation of its upper jaw. Its fins, like those of most scombroids (tunalike fish) are long and narrow. This shape permits steering, but with a minimum of drag through the water. In the sailfish, the pelvic fins are positioned very far forward on the fish's body and are unusually long. Each folds into a groove on the belly when the fish swims at speed.

The tail fin is sickle-shaped, as in many tunalike fish, and provides tremendous thrust to power the fish through the water. The base of the tail, the peduncle, is thin and has two

The sailfish's **first dorsal fin** *is very large and runs most of the length of the body. It is used to corral prey and is also used during courtship displays.*

The boney **"rays"** *of this ray-finned fish are clearly visible.*

The **lateral line** *enables the sailfish to detect changes in water pressure.*

The long and sharply pointed **bill** *is used to slash and stun prey before they are swallowed whole.*

The thin **pelvic fins** *are longer than those of any other billfish.*

pectoral (shoulder) fins

▶ **Sailfish**
The sailfish is remarkable for its sail-like dorsal fin and its long pointed bill. When the sailfish is swimming at speed, the dorsal fin is folded into a groove along the dorsal surface.

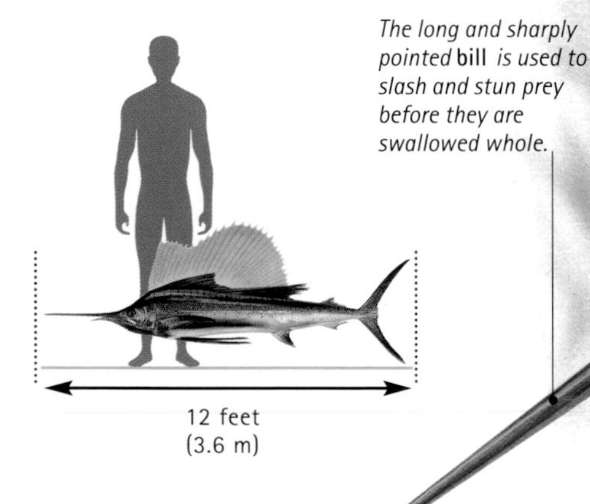

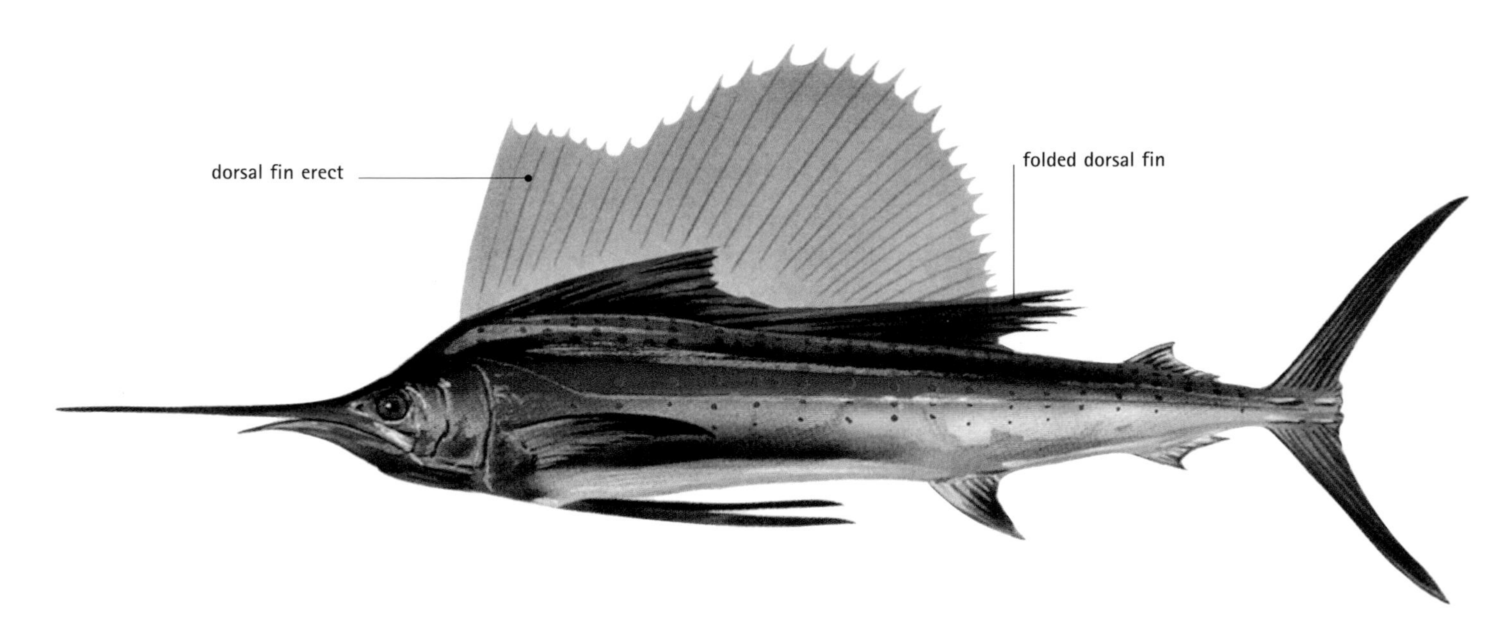

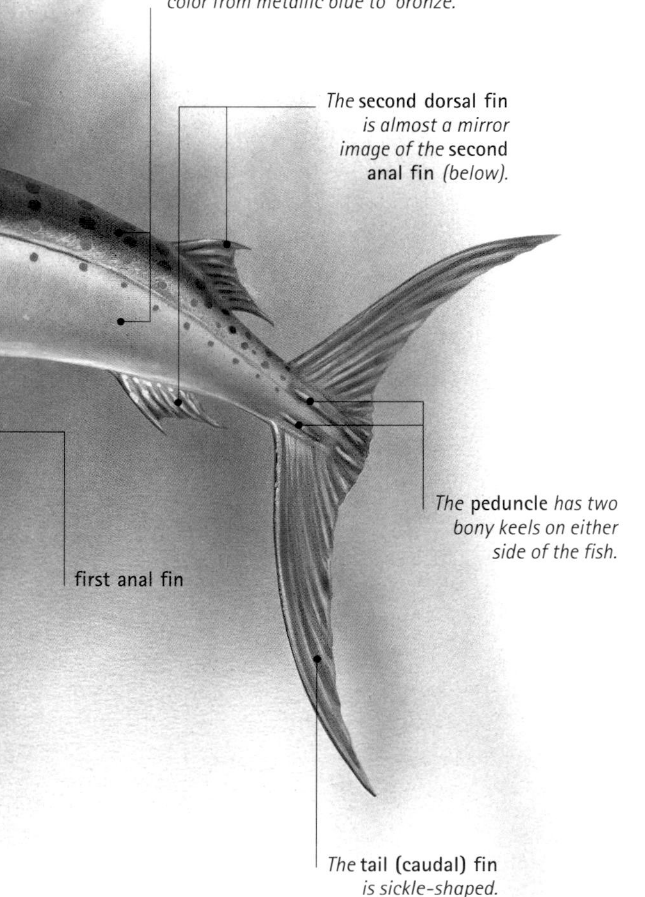

bony keels on either side formed by extensions of the tail vertebrae. These keels help keep the tail in a vertical position and assist the tail in cutting through the water as it moves from side to side during swimming.

The sailfish has two dorsal fins and two anal fins. This species is different from other billfish in the shape and large extent of the first dorsal fin, which may be twice as high as the sailfish's body. This sail-like fin acts as a keel, helping the fish to make fast turns. It also acts as a barrier, keeping prey from escaping over the top of the

▲ ERECT DORSAL FIN
Sailfish
With the sail-like dorsal fin erect, the sailfish can maneuver rapidly and corral prey. With the fin folded down, the sailfish produces less drag and it is able to swim fast.

CLOSE-UP

Countershading

The sailfish, like many other fish that spend most of their time in the sunlit surface waters of the ocean, is countershaded. That is, its body surface is dark on top and pale on the underside, gradually grading between the two. In sailfish, the underside and flanks are white or silvery and the dorsal (upper) surface is dark, becoming paler toward the flanks. This pattern provides good camouflage against the sunlight streaming through the water surface. The upper side, which is sunlit, is darkly shaded; the underside, which is in shadow, is light-colored. The overall effect is to make the fish fairly uniform in brightness when viewed from the side. It is well camouflaged against the sunlit surface waters when viewed from below; and it blends in well against the darker, deeper water when viewed from above. In avoiding predators or seeking prey, the countershading provides good camouflage from whatever direction the sailfish is viewed.

IN FOCUS

Changing conditions

Billfish and tuna spend their lives in the upper layers of warm oceans. However, light intensity, water pressure, and temperature change dramatically over a vertical range of 3,300 feet (1,000 m) or less. At depths below only 660 feet (200 m), light intensity is massively reduced, temperatures are much lower, and pressure is more than 20 times greater that at the surface. Billfish range freely between the surface and depths of about 660 feet (200 m) and so must be able to adapt to such dramatically changing conditions. The swordfish dives more deeply than other billfish. It regularly descends to 1,970 feet (600 m) and more to catch fish, bottom-living crustaceans, and other invertebrates; then it returns to warmer surface waters to digest them.

▼ *The sailfish has a distinctive series of markings on the flanks and first dorsal fin. The steel-blue dorsal fin is covered with black spots, and the iridescent flanks of the body are barred with stripes or rows of spots.*

sailfish's back; and as a scare device. The fin is also held erect by both sexes during courtship displays. The first dorsal fin folds down into a groove along the back for streamlining when the fish is swimming at high speed.

The sailfish is the fastest fish of all: a sailfish hooked by an angler was recorded taking line off his reel at 70 mph (110 km/h). Thus, the sailfish was swimming at about the same speed as the fastest-running animal, the cheetah.

Bill

The bill in sailfish, marlins, spearfish, and swordfish is the long upper jaw. It is covered with tiny toothlike denticles similar to those in the skin of sharks. The spearlike bill is both a weapon and the ultimate in streamlining. It enables the fish to cut through the water at high speed. When sliced back and forth from side to side, the bill can stab or batter prey fish in a school, immobilizing them until the billfish returns to eat them. Among billfish, the bill of the swordfish is the longest, sometimes reaching 5 feet (1.5 m) long and at least half the length of the fish's body. The Mediterranean spearfish has the shortest bill, only one-eighth the length of the body or less.

Color changes

The body color of a sailfish is generally dark steel blue or occasionally bronze above, fading to white or silver below, usually with about 20 vertical bars of pale blue spots along the flanks. The dorsal fin is typically bright blue scattered with black spots of moderate size. Most billfish, including the sailfish, go through spectacular color changes when they attack a school of prey fish. The gill covers, pectoral fins, and flanks turn bright blue, and the vertical stripes along the flanks glow with iridescence. These color changes probably confuse the prey fish, causing them to bunch together so that they are more likely to be stabbed, battered, or slashed by the billfish's swordlike bill.

The scales covering the sailfish's body are small, and a lateral line runs along the midline of both flanks. As in most other bony fish, the lateral line system consists of small canals that open at the body surface through pores. The lateral line system detects vibrations and pressure changes in the water, such as disturbances produced by prey fish.

Skeletal system

Like other bony fish, billfish have a skeleton comprising a skull, jaws, and a backbone, as well as bony spines and rays that support the fins. The component parts of the skeleton are moved by sets of muscles that pull on the bones, which act as levers. In sailfish, the spine, or vertebral column, contains numerous vertebrae that move slightly relative to one another to allow the body to flex. The first half of the vertebral column has points of attachment for ribs that extend around the body cavity to provide some support and physical protection for the soft internal organs contained within. In the head region, the skull provides housing and protection for major organs of vision, smell, and balance. The skeleton surrounding the mouth and gills consists of upper and lower jaws and support for four gills on either side. A bony cover, the operculum, protects each gill cavity.

In sailfish, as in most bony fish, the pectoral fins are attached to the skull by a pectoral girdle; and the dorsal, pelvic, and anal fins are anchored into muscle blocks.

▲ *Like the sailfish, the marlin has a spearlike bill that is an elongation of the upper jawbone.*

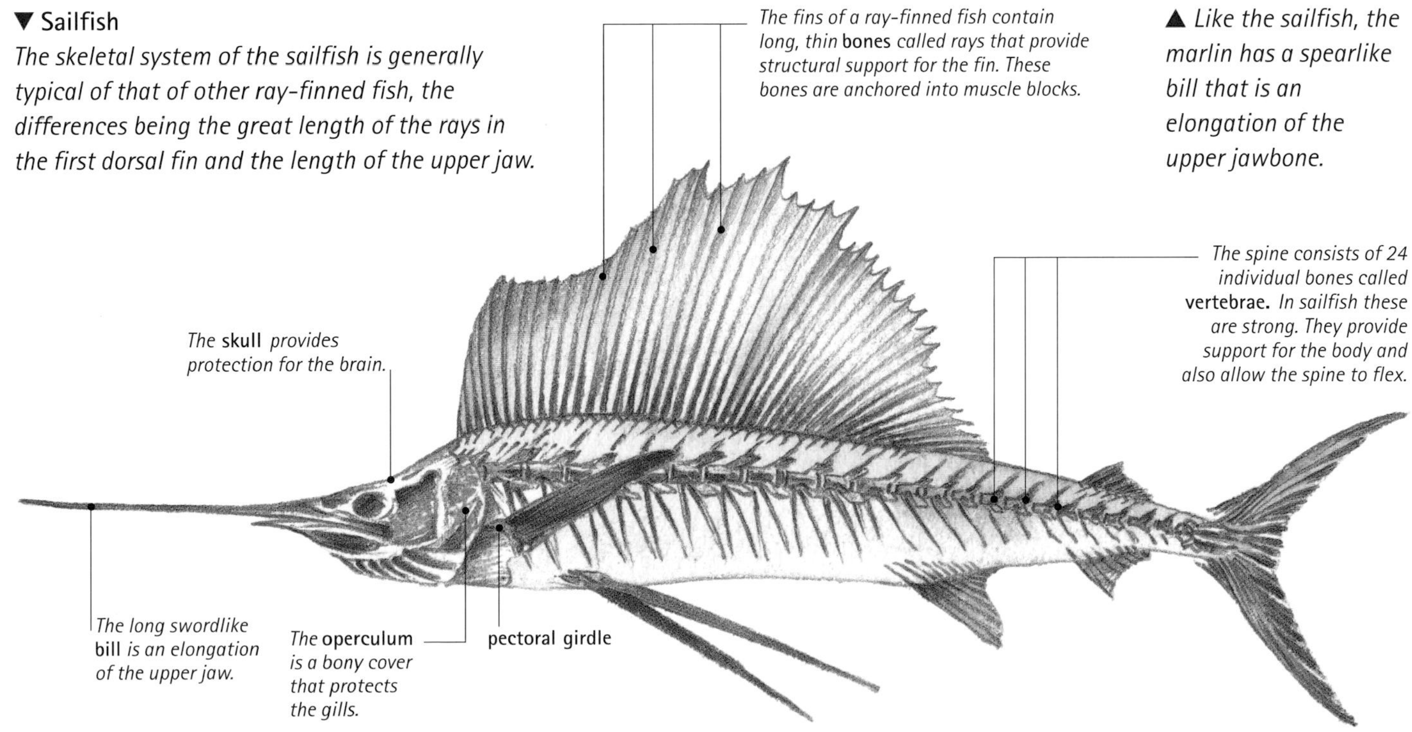

▼ **Sailfish**
The skeletal system of the sailfish is generally typical of that of other ray-finned fish, the differences being the great length of the rays in the first dorsal fin and the length of the upper jaw.

Muscular system

CONNECTIONS

COMPARE the swimming action of a ***TROUT*** with that of a sailfish. Much of the trout's movement occurs as a result of curving the body, but in a sailfish most of the thrust comes from the tail.

Muscles contract to exert power and generate movement. Muscles cannot extend of their own accord, so they are usually arranged in antagonistic sets, with one set working against the other. While one set is contracting, the other set is relaxing and lengthening, ready to contract again.

Streamlined for speed

Fast-cruising fish like tuna and billfish are highly streamlined so that drag (created by water turbulence and friction with the surrounding water) is reduced to a minimum. The pectoral and caudal fins are long but their width, from front to back, is narrow. Such fins are good for forward propulsion at high speed with a minimum of drag. They are not very effective at low speed, because they tend to lack lift and so cause the fish to stall and drop in the water, like a stalling aircraft. Trout and salmon, by contrast, cruise at low speeds but need to accelerate rapidly to catch food. They have much broader fins, with low aspects. Such fins generate more lift and do not stall at low speed, but they produce more drag than high-aspect fins, so they require relatively more muscular effort for fast swimming.

Tail tendons

Billfish and tuna have an unusual mechanism for generating powerful tail strokes. Muscle blocks, called myomeres, on either side of the backbone connect to giant tendons that run backward and over the base of the tail. The tendons are attached to the rays of the tailfin. When the myomeres on one side of the body contract, they pull on the tendons, which run over the base of the tail like a pulley mechanism. This arrangement greatly increases the power of the tail stroke. The muscles and fin rays are also connected so that the tail flexes to produce the maximum forward thrust at all stages of the tail movement. The thrust is like that produced by

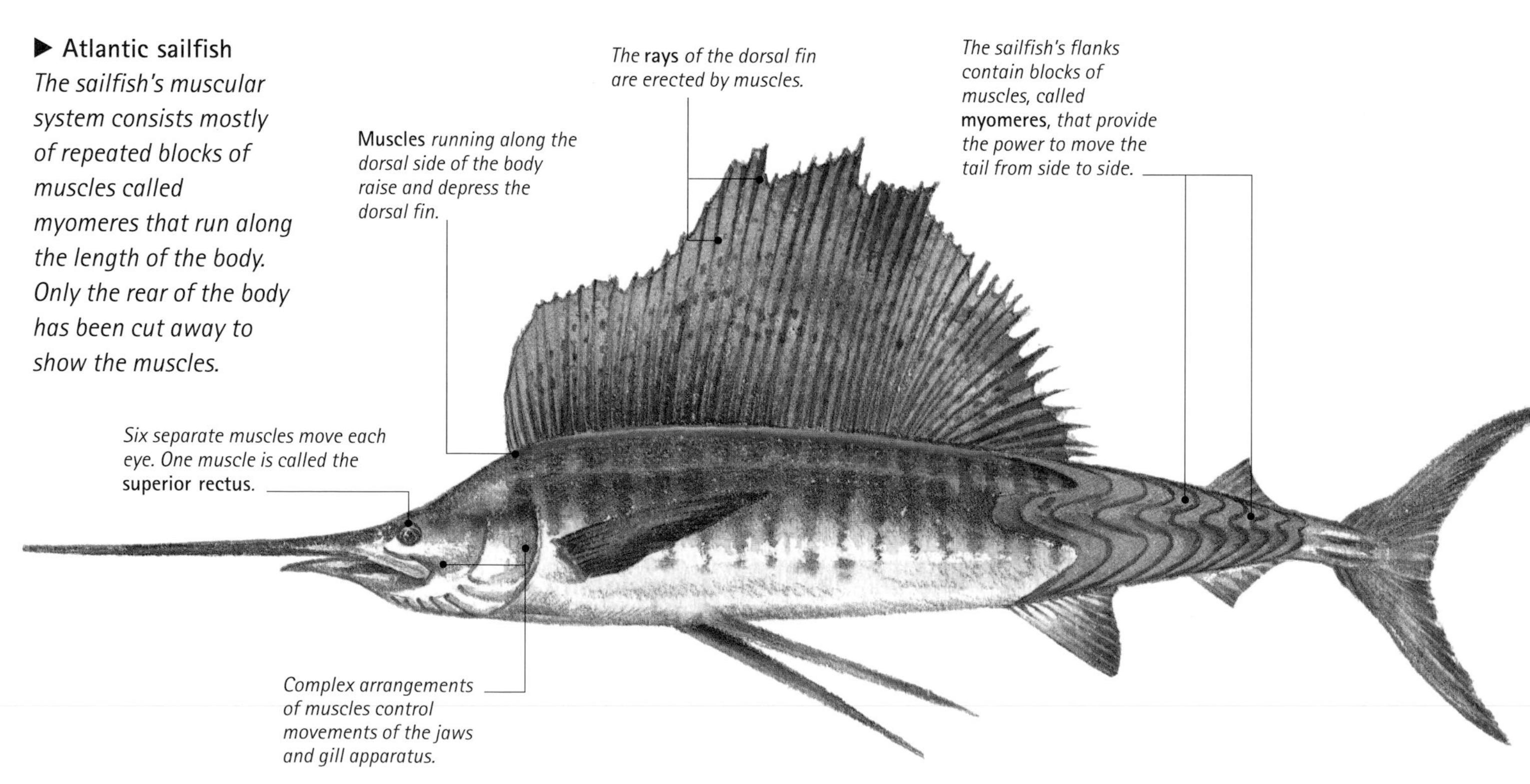

▶ **Atlantic sailfish**
The sailfish's muscular system consists mostly of repeated blocks of muscles called myomeres that run along the length of the body. Only the rear of the body has been cut away to show the muscles.

the propeller of a ship, but back and forth rather than in a circular motion. In billfish and tuna, the tail produces more than 90 percent of the forward thrust in swimming, and the body flexes relatively little. In most other fish, such as trout and salmon, side-to-side movements of the body create much of the forward thrust in swimming.

▲ *The powerful muscles of the yellowfin tuna enable it to swim at speeds up to 40 miles per hour (64 km/h).*

Fish such as tuna and billfish that swim at high speed can cruise at more than five times their body length each second. In most billfish, some of the narrow fins flatten against the body, or fold away into grooves, to reduce drag and turbulence during swimming at the highest speeds. At lower speeds, these fins are extended to reduce rolling.

COMPARATIVE ANATOMY

Red and white muscle

Fast-cruising fish such as billfish and tuna have large regions of red muscle in the swimming muscles along their flanks. These muscles are unusually rich in blood vessels, but the feature that gives them their dark-red color is the presence of myoglobin. This is an oxygen-carrying pigment similar to the hemoglobin found in red blood cells. Myoglobin traps oxygen for use by mitochondria. These are small saclike structures within cells that are specialized for aerobic respiration (using oxygen to break down sugars and fats to release energy). Red muscle cells are rich in mitochondria. These muscle cells require a rich supply of blood to deliver the oxygen and remove carbon dioxide. This enables the muscle cells to respire aerobically to provide the energy for frequent and powerful muscle contractions. The remainder of the flank muscle in billfish and tuna is white tissue. In comparison with red muscle, white muscle has a smaller blood supply, fewer mitochondria, and reduced levels of myoglobin. White muscle provides powerful contractions, resulting in short bursts of high speed. This muscle tires easily, however, so these bursts of high speed cannot be maintained. White muscle can respire without oxygen, but as it does so it releases lactic acid, which soon builds up and causes the muscle to fatigue. In fish such as salmon and trout, which cruise at low speeds but use fast bursts of speed to capture prey, the proportion of red muscle to white muscle is considerably lower than in tuna and billfish.

Nervous system

▲ *Like other billfish, the Pacific blue marlin has large eyes that enable the fish to see in dim light.*

The nervous system of billfish has the same plan as that of other bony fish. The central nervous system (CNS) consists of a brain and spinal cord. The CNS is connected via nerves of the peripheral nervous system (PNS) to sensory organs and to responsive structures, called effectors, such as muscles.

The brain is broadly divided into three regions, as in other vertebrates: the forebrain, hindbrain, and midbrain. Billfish and tuna locate their prey principally by sight and smell. Like tuna, billfish can detect the smell of chemicals, such as oils and proteins, found in the mucous layer covering the skin of prey fish. Like most fish, billfish probably have reasonable hearing and can detect vibrations in the water using their lateral line system. This consists of an extensive system of small canals that open at the body surface through pores along the flanks of the fish. The system detects vibrations and pressure variations in the water that cause the gel-like fluid in the pores and canals to move. This movement triggers sense organs within the canal system, called neuromasts, which send electrical messages to the brain. In that way, the billfish can sense disturbances in its surroundings, such as those produced by an approaching predator or by prey fish.

Large eyes

Recent studies of billfish eyes reveal that most see in color, with the possible exception of the swordfish, which hunts mostly at night. Billfish eyes are unusually large, but this size does not

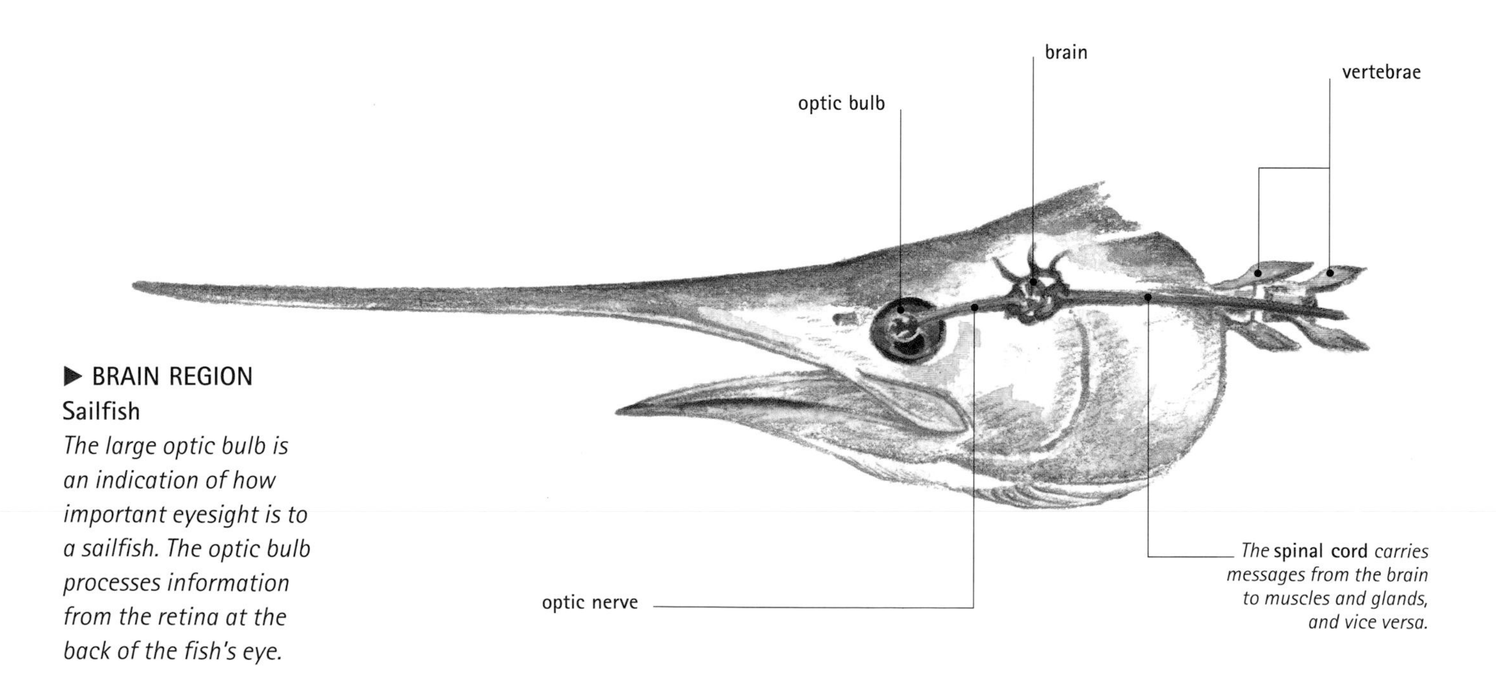

▶ **BRAIN REGION**
Sailfish
The large optic bulb is an indication of how important eyesight is to a sailfish. The optic bulb processes information from the retina at the back of the fish's eye.

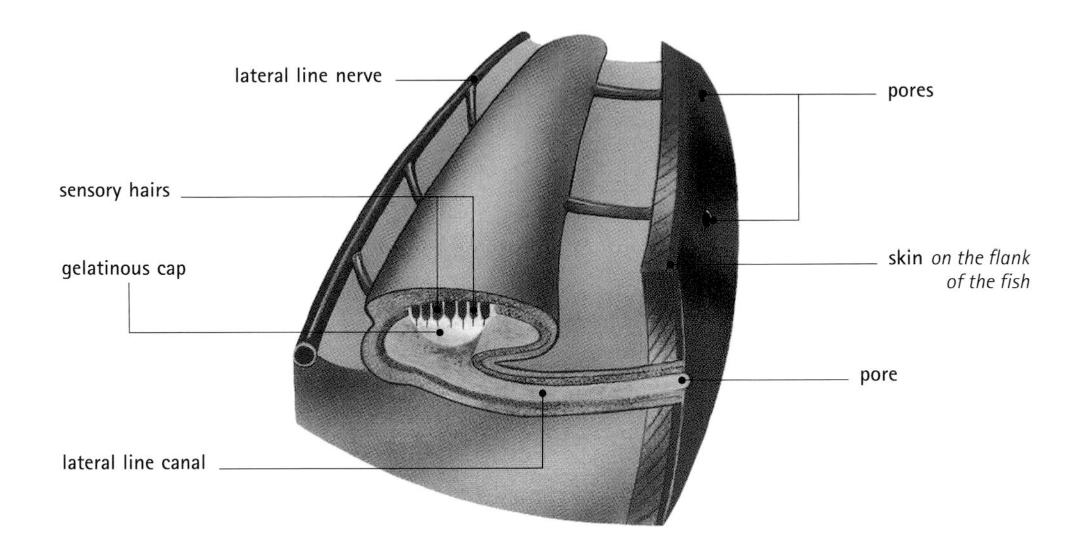

◀ **CROSS SECTION THROUGH LATERAL LINE SYSTEM**
The lateral line is a sensory organ that runs down the flanks and around the head. Sensitive hairs in the lateral line detect movements in the water. Sailfish use their lateral line to detect the swimming of both predators and prey.

give them an advantage in terms of visual acuity—the ability to discriminate small objects. Instead, collections of light-sensitive cells at the back of the billfish's eye connect together, giving the eye great sensitivity to light. Most billfish can see well in dim light and when traveling at high speed. They can see objects that are directly ahead most clearly, but they can also discriminate objects directly behind them without turning the head.

Rods and cones

The light-sensitive region at the back of the eye is called the retina. In most billfish, the upper part of this region receives light coming from the dark depths below the fish. The light-sensitive cells in this region are mostly rods (so named because of their shape); they are good at detecting low levels of light, but not color. The lower region of the retina detects light streaming through the surface waters. The light-sensitive cells in this region include cones (again, so named because of their shape), which detect color but require bright light.

As in other tunalike fish, the eyes of most billfish can adapt to the amount of light available. During daytime, masking pigment helps protect rod cells from bright light, whereas the color-sensitive cones move toward the light. At night, cones move away from the dim light, and the more light-sensitive rods move toward it. These changes ensure that the fish sees well in both the bright light of day and the dim light of night. As in other vertebrates, good color vision is limited to bright conditions only.

Nerve cells and sense organs do not work as quickly as normal if they are chilled. For a fast-swimming predator such as the sailfish, quick reactions are vital if the billfish is to successfully hunt its prey. Billfish have an astonishing mechanism for warming the blood that reaches the brain and eyes to keep them operating efficiently even in cool water. In each eye, one of the muscles, the superior rectus muscle, generates heat. It has a very high rate of respiration and contains numerous mitochondria (small saclike structures inside the cell) that respire using oxygen to generate heat. This heating mechanism, technically called a thermogenic (heat-generating) organ, gives billfish an when hunting in cool water.

COMPARATIVE ANATOMY

Heat organs

Thermogenic organs have evolved quite separately in scombroid (tunalike) fish other than billfish. In tuna, the thermogenic organ is a region of swimming muscle lying close to the backbone. This heat-generating region keeps the tuna's swimming muscles at a temperature well above that of the surrounding water so that the muscles can work more efficiently. The butterfly mackerel has a thermogenic organ similar to that of billfish, but it is a modified lateral rectus eye muscle—not the superior rectus as in billfish.

Tuna, billfish, and butterfly fish are not the only fish that can keep parts of their bodies well above the temperature of their surroundings. Fast-swimming sharks—such as the great white, porbeagle, and mako sharks—keep their muscles warm using a countercurrent heat exchange system similar to that found in dolphins and other cetaceans. Tuna also have such a system, but billfish do not.

Circulatory and respiratory systems

CONNECTIONS

COMPARE the gills of a sailfish with those of a ***TROUT***. The sailfish's gills have a much larger surface area that enables them to achieve maximum gas exchange. The gills also have fused gill filaments, which provide them with added strength.

Like other bony fish, the sailfish has a single-circuit blood circulation, with blood pumped at high pressure from the heart, through the gills, and then on to body tissues before the blood returns to the heart at lower pressure. Major blood vessels called arteries carry blood away from the heart, and major veins return it. Arteries divide into much smaller blood vessels, called capillaries, in the gills and in other body organs and tissues. The capillaries allow gases and dissolved substances to diffuse between the blood and the surrounding body cells.

Four-chamber heart

The heart of a sailfish, like that of most ray-finned fish, contains four chambers in sequence: the sinus venosus, atrium, ventricle, and bulbus arteriosus. Strictly, the atrium and ventricle are true heart tissues, while the sinus venosus and bulbus arteriosus are modified parts of major veins and the main artery, the aorta. The sinus venosus bulges and releases blood smoothly into the atrium at the start of the heartbeat. The atrium then contracts and thus raises blood pressure to supply the ventricle, which is much more muscular and raises blood pressure much higher in the heart's main power stroke. The bulbus arteriosus is an elastic region that dampens changes in blood pressure from the ventricle so that the blood is pumped smoothly around the body, without jerky stops and starts.

▼ *Sailfish swim with their mouth open, allowing water to pass over the gills continuously. This enables oxygen to be taken into the bloodstream and carbon dioxide to be expelled at a high rate.*

CLOSE-UP

Billfish gills

The gills of tuna and billfish do not look quite the same as those of most fish. In tuna and billfish, the stacks of gill filaments that make up a gill are not completely separate, as in most other fish, but are fused in places. Water flows around the connected regions. These connections strengthen the system of gill filaments. In ram ventilation, water is forced at high speed through the gills and the reinforced gill filaments help strengthen the gill against damage from particles that enter. The gills of tuna and billfish often show signs of damage from inhaled objects.

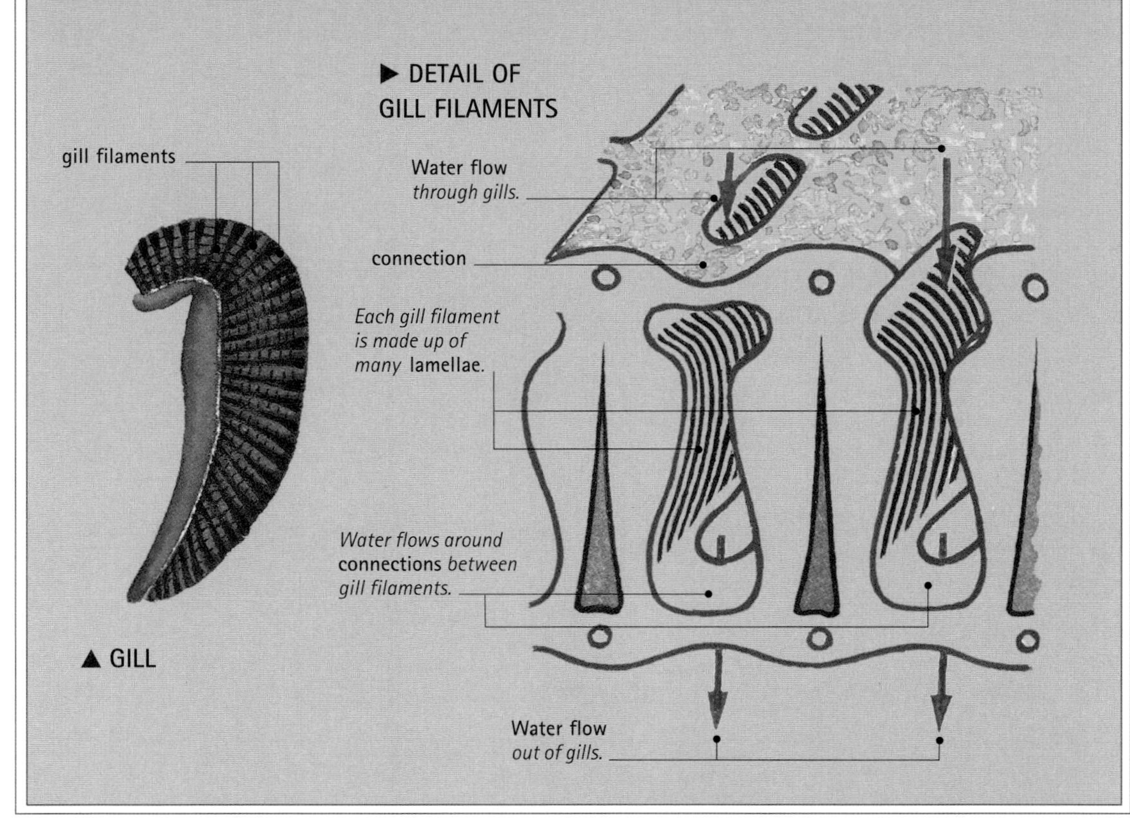

Ram ventilation

Like most other fish, billfish use gas exchange across the gills to gain the oxygen they need for respiration and to get rid of carbon dioxide. To be capable of cruising at high speed, tuna and billfish need to take in oxygen and expel carbon dioxide at a high rate. They do so by swimming with the mouth open all the time, thus forcing a fast stream of water over the gills. This mechanism is called ram ventilation because water is "rammed" through the gill chambers. Most other bony fish pump water over their gills by contracting muscles to change the volume inside the mouth and gill chambers. In comparison, ram ventilation is more energy-efficient, but it does require the fish to swim constantly. If a fish that relies on ram ventilation stops swimming, it suffocates. To permit high rates of gas exchange, the gills of billfish and tuna have a very large surface area—more than 10 times that of most other fish of equivalent size. This enables the tuna or billfish to extract as much as half of the oxygen in the water flowing over the gills.

To deliver the oxygen rapidly to muscle tissue, billfish and tuna have an unusually large heart that pumps the blood at high pressure—commonly at least double the pressure found in other fish and at twice the rate per unit of body mass as in other fish. The volume of oxygen-carrying red blood cells packed into the blood is also unusually high and is at levels similar to those found in diving mammals such as seals and dolphins.

Digestive and excretory systems

CONNECTIONS

COMPARE the large stomach and large intestine of a sailfish with those of another predator, such as a ***LION***. In relation to the sailfish's overall size, the intestine is longer than that of a lion.

Mature billfish do not have teeth but swallow their prey whole without chewing. Sailfish hunt squid and small to medium-size fish such as sardines, anchovies, needlefish, jacks, mackerels, and tuna. When an animal swallows without chewing, consumed food items require considerable physical and chemical digestion to break them down into dissolved food components that can be absorbed. To accomplish this digestion, billfish have a large stomach and, for predators, relatively long intestines.

Swallowed food travels along the esophagus to the stomach, where digestion begins, before the partially digested food passes into the first part of the intestine. The intestine gradually breaks down the various food components—carbohydrates (sugars), proteins, and fats—into smaller chemical substances that are absorbed into the blood. The digested food is delivered to body tissues, where it is used to make body parts or is broken down to release energy in the process of respiration. Any undigested material remaining in the intestines is voided from the body through the anus.

The liver of billfish is large and produces bile that passes into the first part of the intestines, where it helps break down fats. The liver also stores blood sugar in the form of glycogen, which serves as an energy store, and it breaks down toxic substances that are circulating in the blood. The two kidneys in the sailfish's excretory system are also large. As in many other bony fish, the kidneys filter the blood and remove dissolved waste substances that are expelled from the body in urine, which is stored temporarily in the bladder before release.

CLOSE-UP

The swim bladder

The swim bladder of the sailfish, like that of other fish, evolved over millions of years from an outfolding of the gut that has since closed off and become a separate sac. The sailfish swim bladder is unusual in being large and divided into many small chambers. Each chamber is enclosed in a rich network of blood vessels, with a gas gland that releases gases from the bloodstream into the chamber, or absorbs gases from it. This arrangement helps the sailfish adjust its buoyancy very rapidly, enabling it to change depth while swimming at high speed.

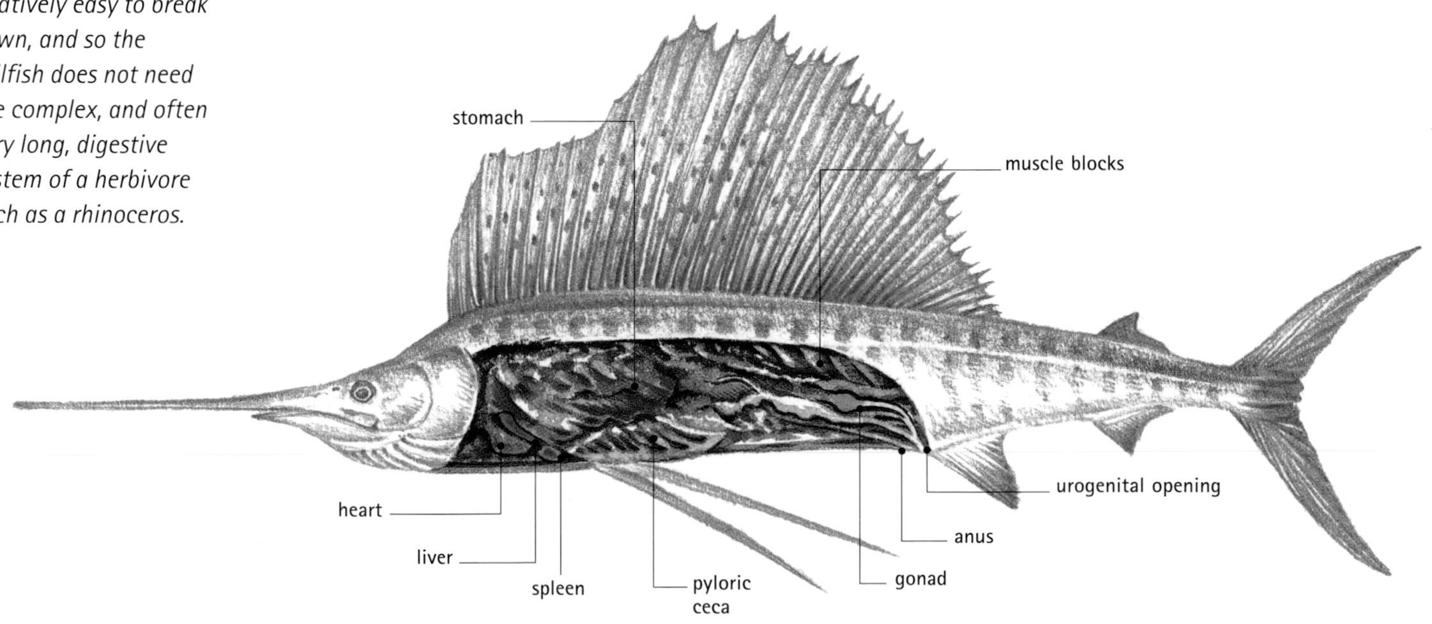

▼ Sailfish

The sailfish's digestive and excretory system is suited to that of a meat eater. Meat is relatively easy to break down, and so the sailfish does not need the complex, and often very long, digestive system of a herbivore such as a rhinoceros.

Reproductive system

Male sailfish have a pair of sperm-producing organs, the testes, in the lower abdomen. Females have two egg-producing organs, the ovaries. At spawning time, females release millions of tiny eggs. These are fertilized in the surrounding seawater by microscopic, tadpolelike sperm, released in the billions by males.

Adult sailfish usually spawn in summer. When ready for spawning, the female swims with her extended dorsal fin above the water surface. She is escorted by one or more males, which jostle with one another other to fertilize her eggs when she releases them. She expels as many as 4.5 million eggs over several hours. The fertilized eggs float in the surface waters for several days and hatch into larvae that are about 0.13 inch (0.3 cm) long. These spiny hatchlings do not resemble adults, but when they are 0.25 inch (0.6 cm) long, their jaws begin to lengthen and by 8 inches (20 cm) they have lost their larval features and look like miniature adults. Sailfish, and other billfish, grow rapidly in the first six months. At the end of this period, sailfish weigh 6 pounds (2.7 kg) and already average 4.5 feet (1.4 m) long. It is during the first six months that most sailfish fall to predators, especially larger fish and seabirds. Sailfish reach maturity after several years, and the oldest can survive to 20 years or more, reach up to 12 feet (3.6 m) long, and weigh up to 275 pounds (125 kg).

TREVOR DAY

▲ *This juvenile sailfish already displays the distinctive dorsal fin.*

IN FOCUS

Migration

Billfish commonly spawn in tropical waters and then return to feeding grounds in warm temperate waters where food is more abundant. Billfish tagged by scientists are regularly recorded making migrations of more than 3,100 miles (5,000 km) between the feeding and breeding grounds. A black marlin set a record for billfish of 6,630 miles (10,680 km) between the waters off Baja California and Norfolk Island in the South Pacific.

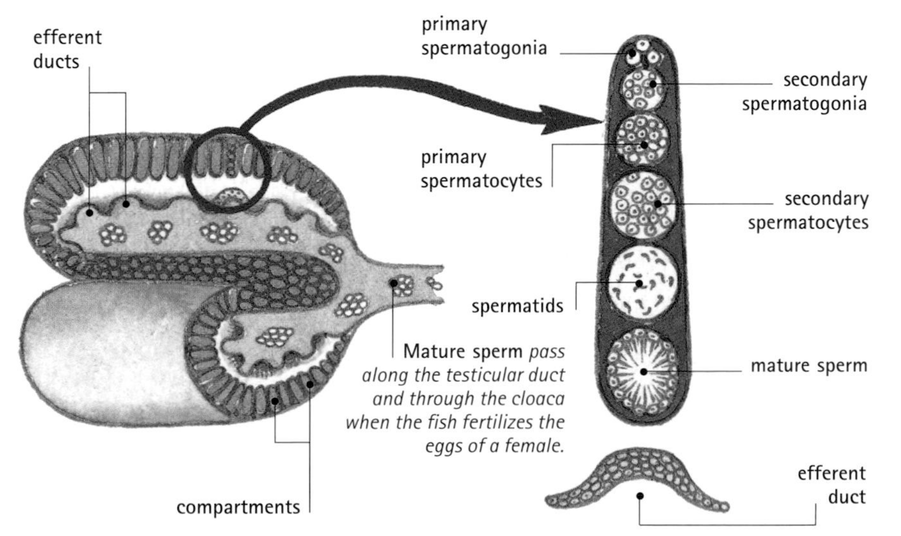

▼ SPERM DEVELOPMENT *Within each compartment of the testis, primary spermatogonia develop through a number of stages until they are mature sperm.*

FURTHER READING AND RESEARCH

Moyle, P. B., and J. J. Cech. 2000. *Fishes: An Introduction to Ichthyology*. (4th ed.) Prentice Hall: Upper Saddle River, NJ.

Nelson, J. S. 1994. *Fishes of the World*. (3rd ed.) Wiley: New York.

Paxton, J. R. and W. N. Eschmeyer (eds). 1998. *Encyclopedia of Fishes*. (2nd ed.) Academic: San Diego, CA.

Scorpion

PHYLUM: Arthropoda CLASS: Arachnida ORDER: Scorpiones

Scorpions are carnivorous arthropods with large claws and a long abdomen ending in a venomous stinger. Scientists have identified about 1,500 different species of scorpions. They occur throughout the world in most terrestrial habitats, from beaches to high mountains.

Anatomy and taxonomy

Scientists called taxonomists organize organisms into groups based on features of their anatomy. Some animals are so similar that microscopic characteristics or even differences in their genetic makeup are used to distinguish between them.

- **Animals** All animals are multicellular and feed on other organisms. They differ from other multicellular life-forms in their ability to move around and respond rapidly to stimuli.

- **Arthropods** These are invertebrates (animals without a backbone) that have jointed legs. Arthropods do not have an internal skeleton but possess a tough outer coating called an exoskeleton that supports their body.

- **Arachnids** Arachnids are almost exclusively terrestrial arthropods. They have two body sections (the prosoma and opisthosoma), unlike insects, which have three. They also differ from insects in having eight legs instead of six and no wings. There are more than 70,000 species of arachnids, and they are classified within 11 orders. Five of these orders (Amblypygi, Palpigrada, Ricinulei, Schizomida, and Uropygi) contain fewer than 100 species.

- **Spiders** Of the remaining six orders of arachnids, the most familiar is the Araneae, which contains more than 35,000 known species of spiders. Spiders feed on other animals and can subdue prey by delivering a venomous bite with two hollow, syringelike fangs.

- **Mites and ticks** The order Acarina contains the mites and ticks and has more than 30,000 known species. Most are very small, with the largest ticks reaching a maximum

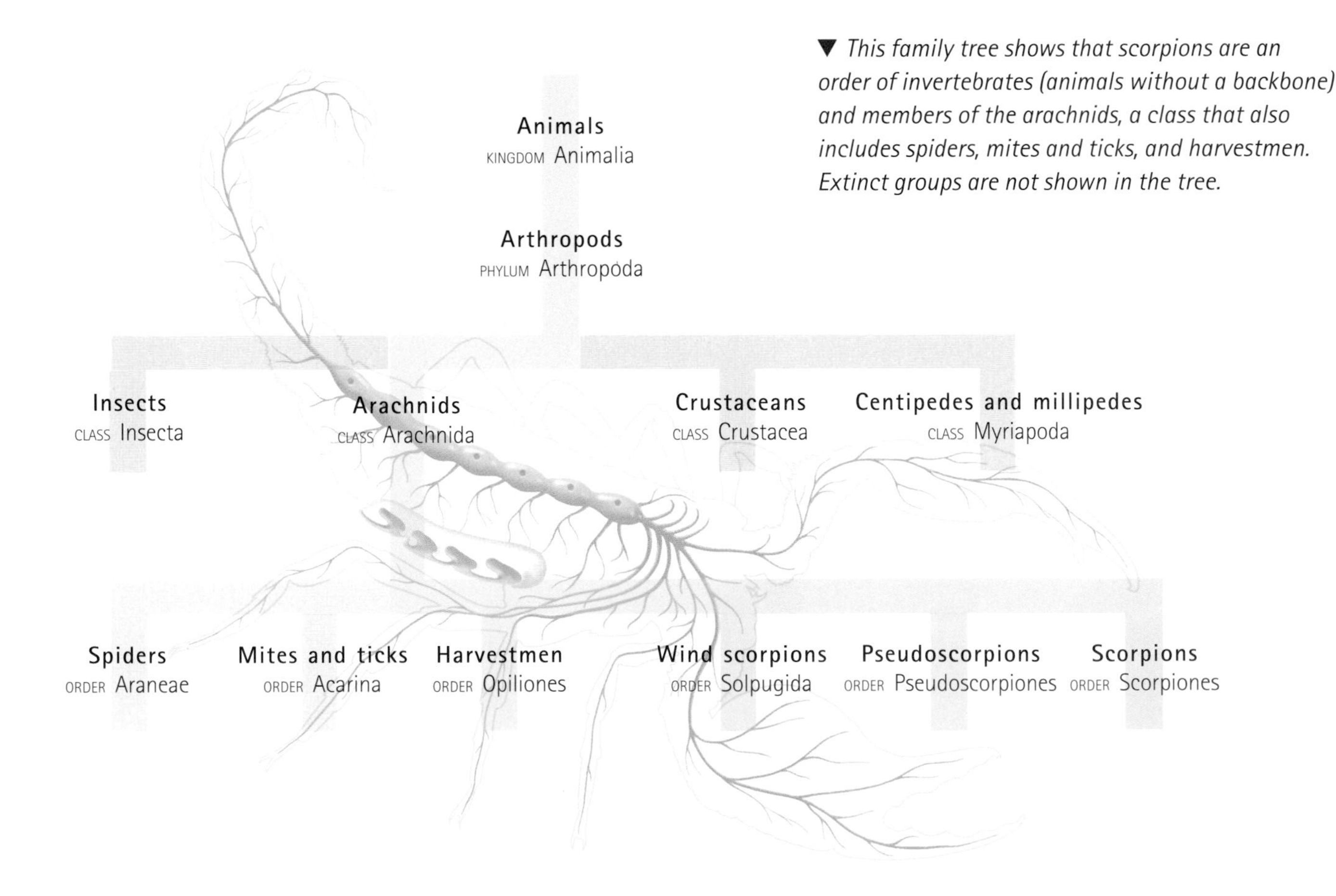

▼ *This family tree shows that scorpions are an order of invertebrates (animals without a backbone) and members of the arachnids, a class that also includes spiders, mites and ticks, and harvestmen. Extinct groups are not shown in the tree.*

length of just over 1 inch (2.5 cm). Most mites feed on dead organic matter. Ticks are external parasites, feeding on the blood of vertebrates. Some ticks need three different hosts to complete their life cycle. While feeding on blood, ticks may transmit disease-causing microorganisms to their host.

• **Harvestmen** These arachnids make up the order Opiliones. Harvestmen are superficially similar to spiders, but their two body parts are fused rather than being clearly defined as in spiders. Unlike spiders, harvestmen cannot produce silk and do not possess poison glands. Harvestmen are omnivores—animals that eat both plants and animals—but some species also eat dead organic matter.

• **Wind scorpions** Arachnids in the order Solpugida are called solpugids, or sun or wind scorpions. They live in tropical and semitropical regions and are extremely active and fast-moving predators. Solifugids run using their hind three pairs of legs; the first pair of legs has a sensory role, detecting obstacles (or prey). Wind scorpions have two pairs of appendages on their head: pedipalps and chelicerae. The pedipalps are large and end in a sticky pad used to capture and draw prey toward the head; and the pincerlike chelicerae are used to kill and dismember prey.

▲ *A black scorpion, native to southern Africa, eats a grasshopper. The scorpion uses its powerful pincers to hold the prey.*

FEATURED SYSTEMS

EXTERNAL ANATOMY Scorpions have two parts to their body: the prosoma (fused head and thorax) and the opisthosoma (abdomen). They have four pairs of legs, two appendages called pedipalps that end with pincers, and a long abdomen with a sting. *See pages 1076–1077.*

MUSCULAR SYSTEM To move their limbs, scorpions use their muscles, the elastic properties of the exoskeleton, and the pressure of their body fluids. *See page 1078.*

NERVOUS SYSTEM Although most scorpions possess eyes, their vision is poor. Instead, scorpions rely on hairs that cover their body to detect obstacles and vibrations. They also have sense organs called pectines. *See page 1079.*

CIRCULATORY AND RESPIRATORY SYSTEMS The scorpion's body cavity contains a fluid called hemolymph. Scorpions breathe using book lungs. The heart pumps hemolymph into the body cavity and around the book lungs, where gas exchange takes place. *See page 1080.*

DIGESTIVE AND EXCRETORY SYSTEMS Food is partially digested outside the body of the scorpion. Exuded digestive fluid breaks down the food. Hairs around the scorpion's mouth prevent the entry of indigestible material. *See page 1081.*

REPRODUCTIVE SYSTEM Male and female scorpions perform an elaborate courtship dance. Scorpions give birth to live young that climb onto the back of their mother after birth. *See pages 1082–1083.*

• **Pseudoscorpions** These are small arachnids in the order Pseudoscorpiones. The largest pseudoscorpion reaches 0.3 inch (8 mm) long. Pseudoscorpions have pedipalps that are pincers, a feature they share with true scorpions. Pseudoscorpions have no tail or sting, but they do have venom glands in their pedipalps. They feed on small invertebrates that they catch with their pedipalps. They live in most habitats. Sometimes, pseudoscorpions make use of another animal for transportation: they grab a passing insect with their pedipalps and "hitch a ride" to another location. This behavior is called phoresy.

• **Scorpions** The true scorpions (order Scorpiones) have two distinctive features that make them easily distinguishable from other arachnids. True scorpions have pedipalps in the form of pincers, and the posterior (rear) five segments of the abdomen are long, forming a tail that ends in a stinger. The tail and sting are curled forward over the body of the scorpion, and the stinger is used both for defense and to capture prey. Prey are usually grasped by the pincers, and the stinger is then used to kill or paralyze. Scorpions live mostly in tropical and semitropical regions, and some species can survive in both desert and humid environments. The order is divided into 14 families. Some of the largest scorpions belong to the genus *Pandinus* (family Scorpionidae). For example, *Pandinus imperator*, which lives in west Africa, grows to 7 inches (18 cm) long.

External anatomy

CONNECTIONS

COMPARE the pedipalps of the scorpion with those of the ***TARANTULA***. Male and female scorpions have pedipalps that end in a pincer. Female tarantulas have pedipalps that look like small legs, but the pedipalps of the male end in a structure called a palpal bulb, which is used to transfer sperm to the female during mating.

Scorpions have a slender body, four pairs of jointed legs, long pedipalps with pincers, and a tail with a stinger at the end. The final part of the abdomen is often curved forward over the scorpion in a characteristic pose.

The body is divided into two sections: the prosoma, or cephalothorax (fused head and thorax); and the opisthosoma (abdomen). The prosoma is smooth and not clearly segmented. In comparison, the scorpion's abdomen is highly segmented and separated into two regions. The front seven segments of the abdomen are called the mesosoma. The next five segments form the thinner metasoma, or tail, of the scorpion. At the end of the metasoma is the sting, or telson.

The scorpion has an exoskeleton—a tough outer layer, or cuticle, that covers the entire body. Like all arthropods, scorpions lack an internal skeleton, and support is provided

▲ *The scorpion's stinger, or telson, is at the end of its tail. The stinger is composed of two sections: the clawlike barb, or aculeus; and a bulbous venom-containing vesicle.*

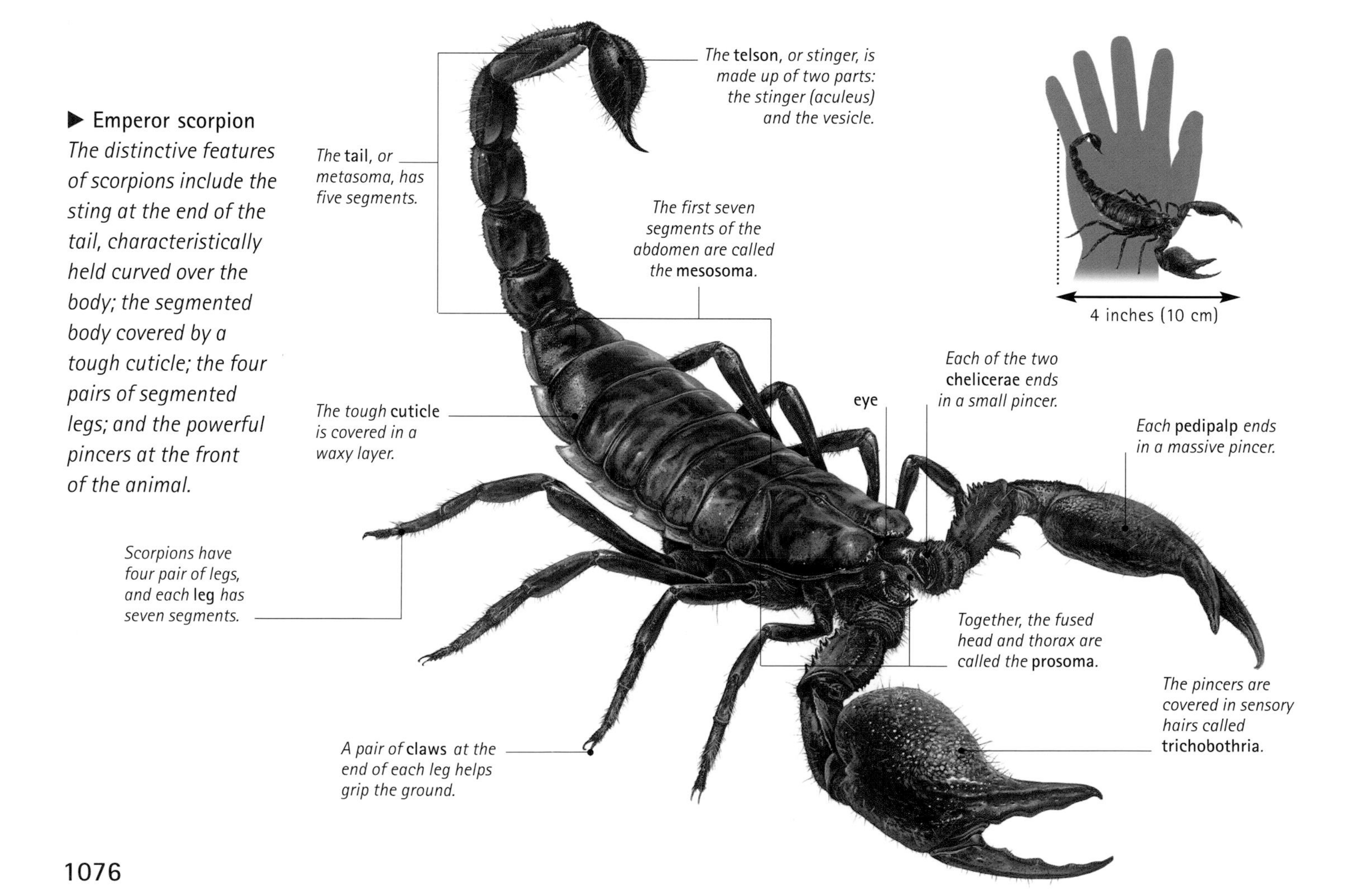

▶ **Emperor scorpion**
The distinctive features of scorpions include the sting at the end of the tail, characteristically held curved over the body; the segmented body covered by a tough cuticle; the four pairs of segmented legs; and the powerful pincers at the front of the animal.

solely by the tough cuticle. The cuticle also has a thick waxy layer that prevents water loss and offers some protection.

Scorpions have eight legs. Each leg comprises seven segments and ends in a pair of small claws that helps grip the ground. The legs are used primarily for walking, but some species also use their legs to dig burrows. Scorpions have a pair of large pedipalps that join the prosoma at the head of the animal. The pedipalps end in a pincer, or chela, that is used to grab prey, for defense, and during courtship. The pedipalps are covered with sensory hairs called trichobothria; these are tactile organs that give the scorpion a sense of touch. As a scorpion walks, it usually holds its pedipalps to the front so the trichobothria are able to sense obstacles, predators, and prey. The trichobothria are the scorpion's main sense organs, but many species also have simple eyes. The number of eyes varies among species but can be as many as six. There is usually one pair of eyes in the center of the prosoma, with others on the front edge of the prosoma.

As well as the tactile hairs and simple eyes, scorpions have sense organs called pectines, which are situated beneath the scorpion behind the most posterior pair of legs. Pectines are gill-like, V-shape organs, and the two arms of the V may contain as many as 50 serrations, like the teeth of a comb. These sense organs are not fully understood, but they are thought be involved in detecting pheromones (chemical signals) and vibrations on the ground.

EVOLUTION

Giant ancestors?

Some scientists believe that modern scorpions evolved from giant marine arthropods called eurypterids. These ancient arthropods lived in the oceans more than 400 million years ago. Some grew to more than 6.5 feet (2 m) long. Fossilized eurypterids have some features that are similar to modern scorpions, including small chelicerae, front appendages with spines for grasping prey, and a long pointed tail.

Each chelicera—at the front of the head—has three segments. The chelicerae end in small pincerlike structures and are used primarily for feeding. The scorpion's telson, or stinger, has two parts. The pointed sting is called the aculeus, and the larger part is called the vesicle. Even though many people fear scorpions, only about 30 species (out of a total of 1,500 known species) are dangerous to humans.

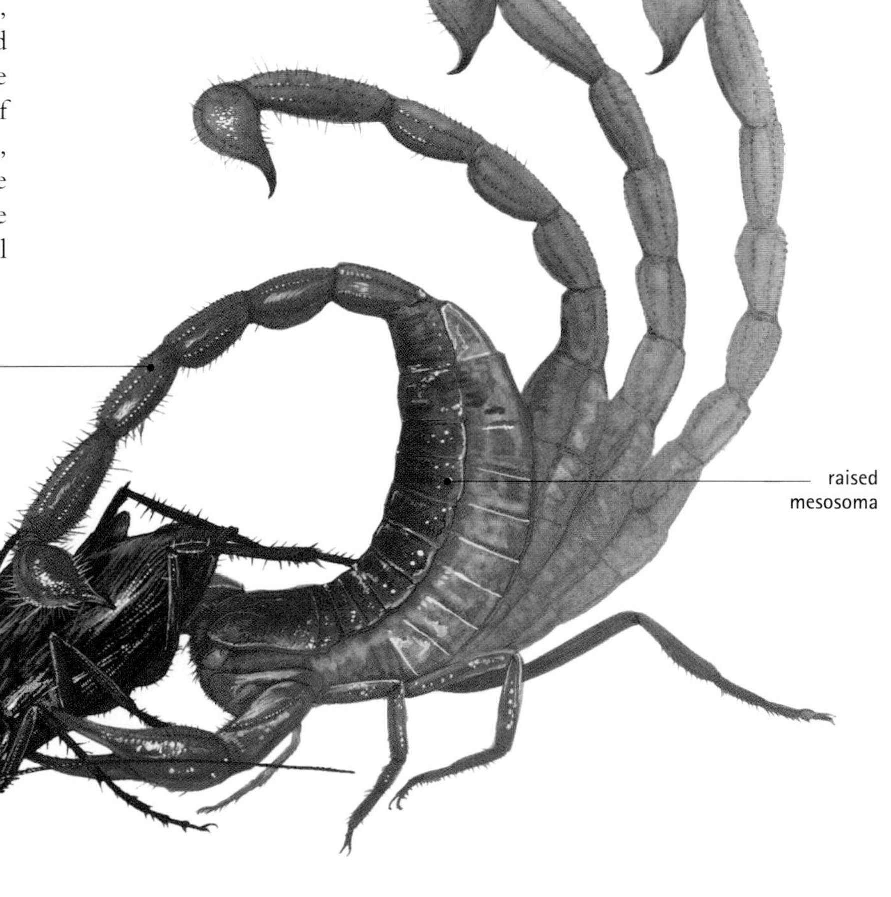

▶ STINGING PREY
Emperor scorpion
When prey, usually an insect, is detected, the scorpion grasps the prey in its pincers. If the pincers alone cannot subdue the prey, the scorpion curves its abdomen over its body to stab the struggling prey, injecting venom through the stinging barb. The sequence of stinging takes about 0.75 second.

Muscular system

Scorpions move their appendages, such as the legs and pedipalps, using muscles. The scorpion's exoskeleton has internal extensions called apodemes, and these provide points for muscles to attach. Muscles within the prosoma extend into the various segments of the legs and pedipalps, and muscles also run the length of the abdomen, connecting each of the segments.

When walking, scorpions typically move two legs on each side of their body at roughly the same time. In this way, four legs may be off the ground at any one time. The fourth leg on one side of the body moves first, then the third leg on the opposite side of the body. It is followed by the second leg on the same side as the first leg to move, and finally the front leg on the opposite side moves. The process is then repeated for the remaining legs—but starting with the fourth leg on the opposite side.

Muscles are also used in breathing. They open small pores called spiracles, allowing air into the chambers that contain the scorpion's four respiratory organs, or book lungs. Scorpions need to be able to control the opening of the spiracles to prevent water loss. All scorpions, particularly species that live in the desert, must conserve as much fluid as possible. If the spiracles stayed open all the time, the scorpion would desiccate (dry out) and die.

CLOSE-UP

Pedipalp power

The pedipalps of some scorpions are very large and packed with muscle. The pincers at the end of the pedipalps of an emperor scorpion are strong enough to break human skin and draw blood. However, it is not only muscles that control the movement of the pedipalps and pincers: scientists have discovered that the cuticle has some elastic properties that allow the pincers to open and the pedipalps to extend without the use of muscles. Some scorpions are also able to enhance this movement by increasing the pressure of their body fluids.

▼ *The powerful pincers at the end of the scorpion's pedipalps, as in the wind scorpion below, are large and packed with muscle.*

Nervous system

Scorpions have a brain in the front end of the prosoma. The brain is connected to two ganglia (singular, ganglion). These are swellings of nerve tissue that contain groups of nerve cell bodies. The ganglia are located above and below the esophagus, the tube that connects the mouth to the stomach. From the lower ganglion, a pair of nerve cords extends the length of the body. The nerve cord consists of a number of ganglia, usually one for each body segment. Nerve fibers branch and extend from the nerve cord into the limbs and body of the scorpion. The nervous system receives impulses from the sense organs, processes the information, and produces a response.

The main sense organs are the trichobothria and the pectines. The trichobothria are touch-sensitive hairs that cover the pedipalps. They can also detect vibrations and changes in air pressure that might indicate nearby predators or prey. Scientists believe that one function of the pectines is to detect chemicals known as pheromones produced by other scorpions. Pectines may also be used to detect vibrations underground or changes in the ground surface.

Simple eyes

Unlike insects, such as ants and bees, scorpions do not have compound eyes; in fact, some cave-dwelling species do not have eyes at all. Other scorpions do possess simple eyes, and these allow scorpions to differentiate between light and dark. Behavioral experiments have shown that some scorpion species prefer to move to dark areas rather than stay in well-lit areas. In addition, some species of scorpions have been found to have light-detecting organs in their tail. The function of these organs is not known, but some scientists believe they are used when the scorpion is headfirst in a burrow so it can still determine the light levels outside.

▼ Emperor scorpion

A scorpion's nervous system includes a brain connected to two ganglia, which are situated above and below the esophagus. A pair of nerve cords arises from the lower ganglion and runs the length of the body. Small nerves branch off from the brain and ganglia.

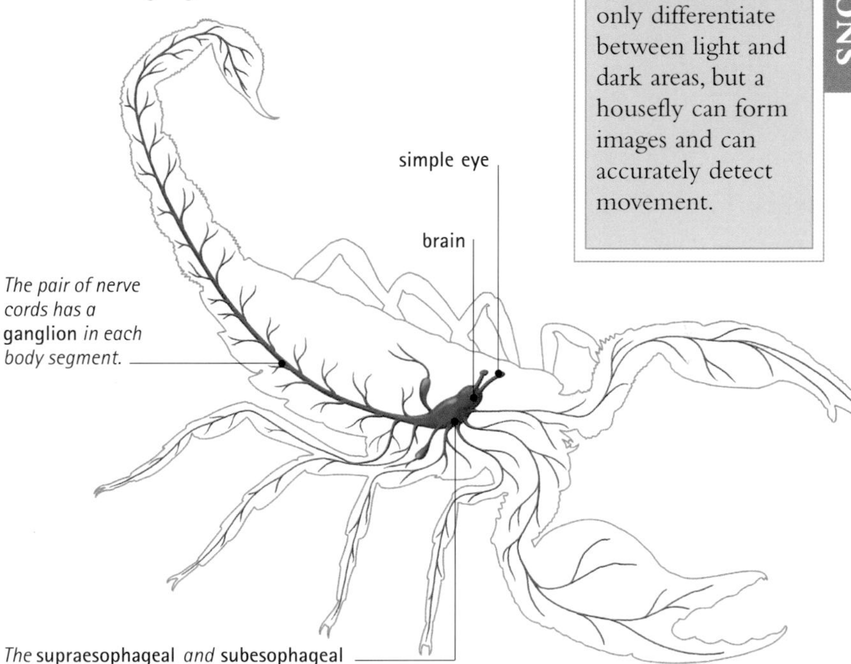

CONNECTIONS

COMPARE the simple eyes of a scorpion with the more complicated compound eyes of a ***HOUSEFLY***. The scorpion can only differentiate between light and dark areas, but a housefly can form images and can accurately detect movement.

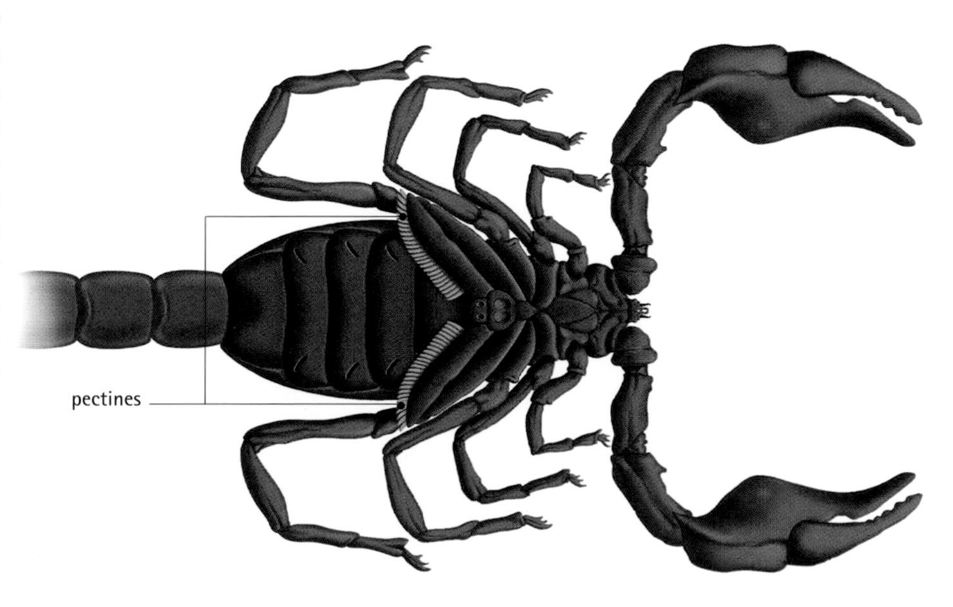

▶ PECTINES

On the underside of scorpions are two gill-like serrated sense organs called pectines. They are thought to be involved in dectecting vibrations and chemical signals called pheromones.

Circulatory and respiratory systems

CONNECTIONS

COMPARE the "open" circulatory system of the scorpion with the "closed" system of the ***CHIMPANZEE.*** The chimpanzee has vessels of different sizes that enclose the blood at all times.

Scorpions have a tubular heart running the length of the mesosoma. At the end closest to the prosoma (the head region), the heart is connected to the anterior aorta, which extends into the prosoma. The other end of the heart is connected to the posterior aorta, which extends into the metasoma (tail).

A bloodlike fluid called hemolymph enters the heart through small openings called ostia. There is one pair of ostia in each of the seven segments of the mesosoma. Heart muscles in the heart wall cause it to constrict, forcing the hemolymph forward and backward into the anterior and posterior aortas respectively. Thinner vessels that lead into the limbs branch off the anterior aorta. These vessels eventually open into the hemolymph-filled body cavity, called the hemocoel. The scorpion's internal organs lie within the hemocoel, and this "open" circulatory system allows hemolymph to flow around the organs. Hemolymph does not contain blood cells used for respiration, but it does contain proteins that carry oxygen and nutrients to the organs and carry wastes such as the gas carbon dioxide away.

CLOSE-UP

Respiration

Scorpions respire through four pairs of book lungs, one on each side of the third to sixth segments of the mesosoma. Each book lung lies within a small chamber, and the external entrance is controlled by a spiracle. Book lungs consist of stacks of platelike structures called lamellae. The lamellae give the book lungs their name because they look like closed pages of a book. The inner side of the lamellae are bathed in hemolymph in the "open" circulatory system. When the spiracle is open, air flows into the chamber containing the book lung. Oxygen diffuses through the lamellae and into the hemolymph.

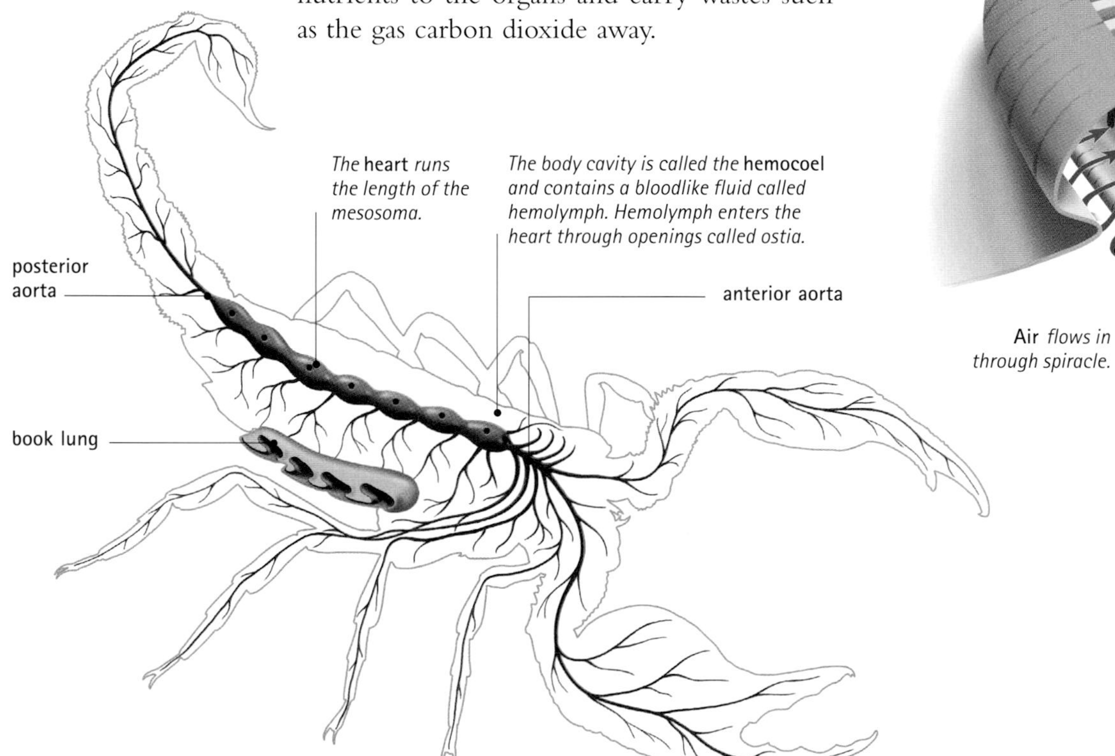

▼ Emperor scorpion
The heart pumps hemolymph around the body. Hemolymph returns to a body cavity called the hemocoel, where it bathes internal organs and picks up oxygen from the book lungs.

Hemolymph *flows between lamellae.*

lamella of book lung

Air *flows in through spiracle.*

▲ BOOK LUNG
Scorpions respire using book lungs. Oxygen passes from the air across the platelike lamellae into the hemolymph. Waste carbon dioxide moves the other way.

Digestive and excretory systems

Scorpions feed mostly on other arthropods. However, some larger scorpions are known to feed on small mammals, amphibians, and lizards. An item of prey is grasped with the pedipalps and paralyzed or killed by the sting. The prey is then pulled up against the head region, or prosoma. The segmented chelicerae then break up the prey into smaller, manageable pieces. The scorpion exudes digestive fluid onto the prey, breaking down its tissues. The partially digested food is then sucked into the mouth and down the esophagus to the stomach. The bristles surrounding the mouth keep indigestible pieces of the prey from being swallowed.

Once in the stomach, the food is broken down further by the action of other digestive fluids. The food then passes into the midgut and intestine, where the nutrients are absorbed.

IN FOCUS

Scorpion venom

All scorpions have a venomous sting, but the toxicity of the venom varies among species. Although the sting is used in defense, it is also important when a scorpion is trying to catch prey. If the venom from the sting quickly kills or paralyzes a scorpion's prey, then the prey will not be able to escape or damage the scorpion during a prolonged struggle. In general, scorpions that have large pincers rely on them to overpower prey and have less venomous stings than scorpions with small, slender pincers; the latter often rely on a potent sting. Some of the most dangerous species—for example those in the genus *Androctonus*, such as the fat-tailed scorpion—can kill a human within a few hours of stinging. However, scorpion antivenin is available in many regions where scorpions live, and it has reduced the number of human deaths.

Waste products pass down the intestine, which extends the length of the metasoma; and are expelled through the anus, which opens in the final segment of the tail (the last segment before the telson).

Scorpions also have excretory organs called coxal glands that excrete the metabolic waste uric acid. The opening of the coxal glands is located at the point where the third pair of legs joins the prosoma.

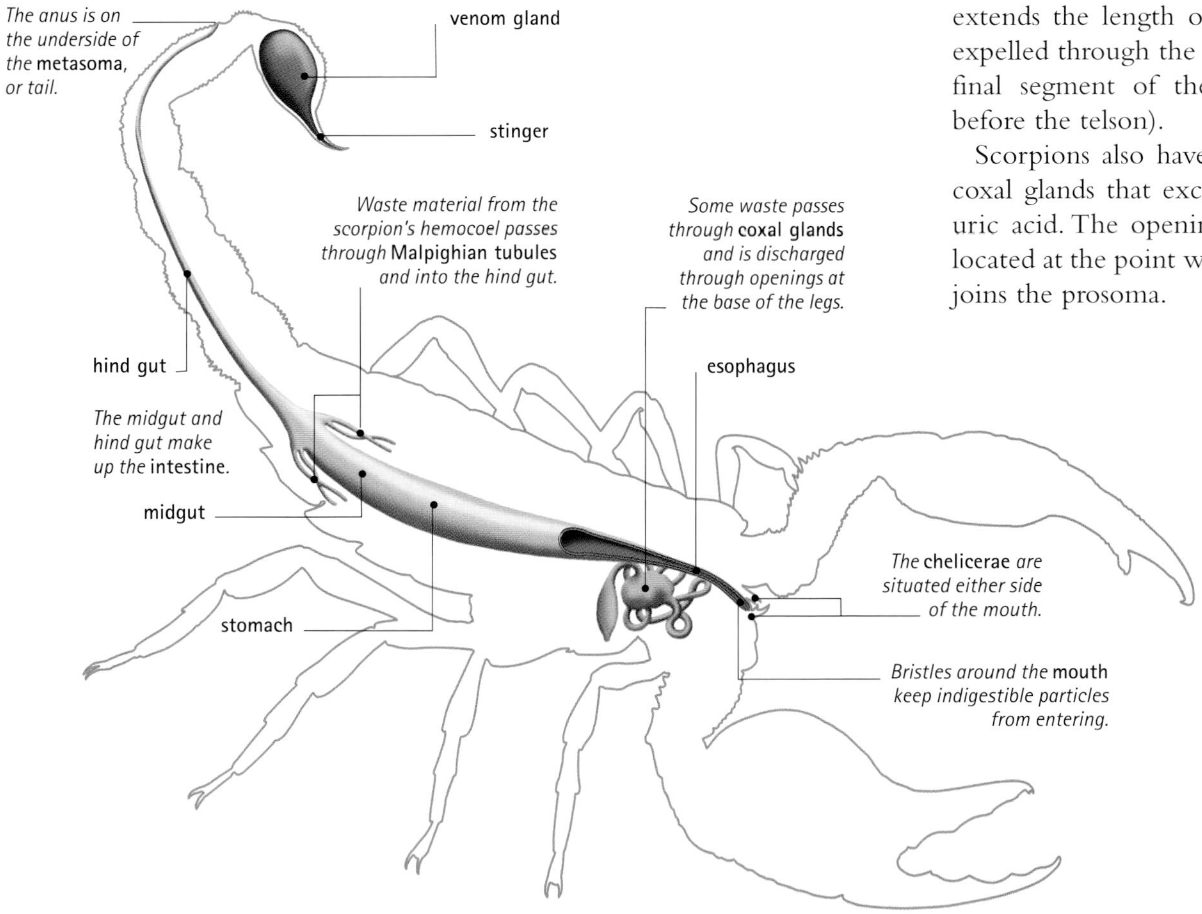

The anus is on the underside of the metasoma, *or tail.*

venom gland

stinger

Waste material from the scorpion's hemocoel passes through Malpighian tubules *and into the hind gut.*

Some waste passes through coxal glands *and is discharged through openings at the base of the legs.*

hind gut

esophagus

The midgut and hind gut make up the intestine.

midgut

stomach

The chelicerae *are situated either side of the mouth.*

Bristles around the mouth *keep indigestible particles from entering.*

◀ Emperor scorpion
The scorpion exudes a digestive fluid that breaks down the tissues of prey items. Partially digested food is then eaten, digested further, and absorbed as it passes through the digestive tract. The metabolic waste uric acid is excreted by the coxal glands.

Reproductive system

CONNECTIONS

COMPARE the courtship dance in scorpions with mating in ***DRAGONFLIES***. The male scorpion deposits a spermatophore on the ground and then positions the female over the spermatophore, which she opens so sperm can enter her genital opening. Dragonflies assume the "wheel" posture, in which the male transfers sperm directly into the female.

Male and female scorpions probably find mates by using chemical signals called pheromones. Possibly, scorpions use their pectines (sensory organs on the underside) to detect pheromones. The scorpions then follow the pheromone trail to find a mate. Once a male and a female scorpion meet, the male grasps the female's pedipalps with his pedipalps and leads her in an elaborate dance called the *promenade à deux*. During this dance the male and female occasionally "kiss" with their chelicerae.

At the end of the dance, the male finds a flat piece of ground on which he deposits a package of sperm, called a spermatophore. He then leads the female over the top of the spermatophore. The spermatophore has an opening lever, and the female presses down on this, releasing sperm into her reproductive tract.

The female reproductive system consists of up to four tubes, depending on the species, running along the mesosoma. The tubes are connected to the genital opening at the front of the mesosoma and are interconnected by other tubes running laterally. In scorpions, embryos develop in the mother in either of two ways, depending on the species. In some species, the embryos absorb nutrients from the surrounding tissues, and the mouth and digestive system develop late in gestation. This process is called apoikogenic development and is also found in some marine

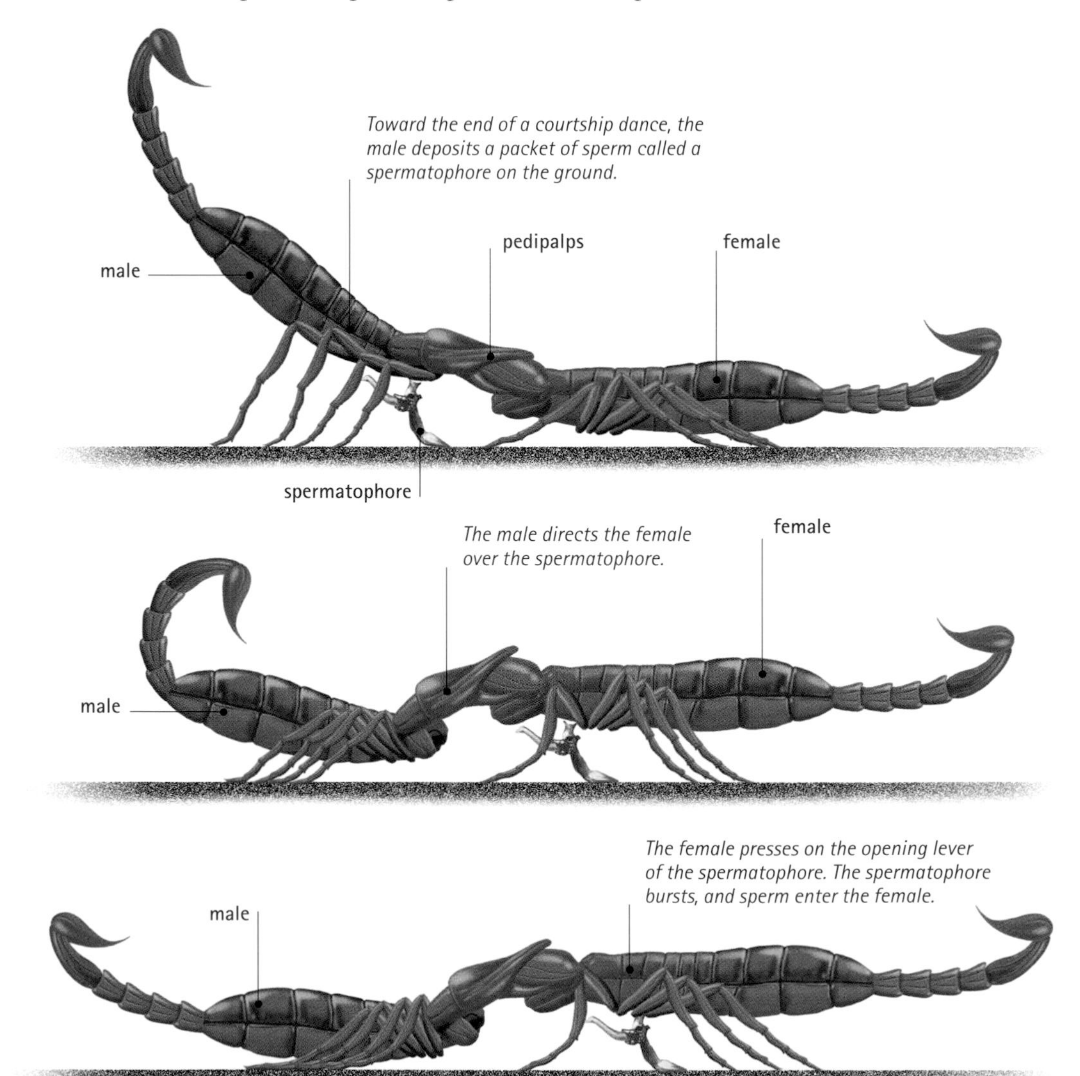

▶ **MATING DANCE**
Emperor scorpion
The male holds the female's pedipalps in his own, and deposits a spermatophore (packet of sperm) on the ground. He then pulls the female over the spermatophore, which she opens, and sperm enter her genital opening, or gonopore.

▲ *Female scorpions are oviparous—they give birth to live young. The juveniles live on the mother's back for two to three weeks, until after their first molt.*

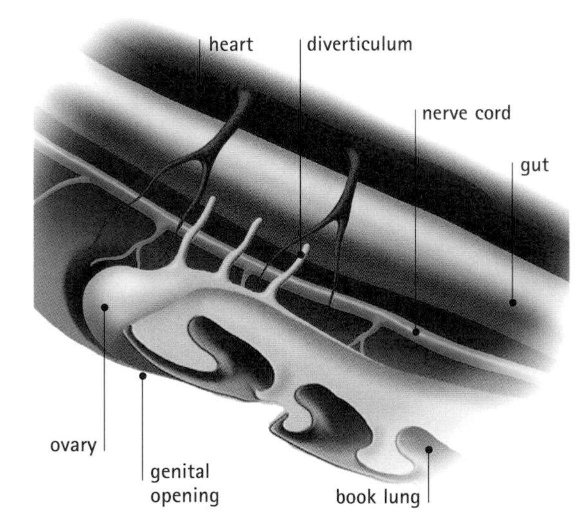

▶ FEMALE REPRODUCTIVE SYSTEM
Emperor scorpion
Embryos develop in tubes connected to the mother's genital opening. In emperor scorpions, each embryo is nourished by a diverticulum, an extension of the gut that connects with the embryo's mouth.

arthropods. In other species of scorpions, the embryos develop the mouth and digestive tract early in gestation and then receive nourishment from the mother scorpion by way of a feeding structure called a diverticulum, which connects the mouth of the embryo and the gut of the mother. This is called katoikogenic development.

In the emperor scorpion, gestation lasts between seven and nine months and, like all scorpions, females give birth to live young. When the young are due to emerge from the female's genital opening she lifts herself up so that her first and second pairs of legs form a cradle beneath her body. As the young emerge from the genital opening, they climb up the female's legs and onto her back, where they remain until after their first molt. During this time, the mother protects her offspring from predators, and the young probably absorb fluid from the cuticle of the mother. Young scorpions have a body shape similar to that of adults and must molt their cuticle to grow.

After the first molt, which can occur days or weeks after birth, depending on the species, the young scorpions leave their mother's back and become independent.

KIEREN PITTS

FURTHER READING AND RESEARCH

Insects and Spiders of the World. 2003. Marshall Cavendish: Tarrytown, NY.

Ruppert, Edward E., and Robert D. Barnes. 1994. *Invertebrate Zoology.* Saunders College: Fort Worth, TX.

Vaejovis carolinianus (Carolina scorpion): www.lander.edu/rsfox/310vaejovisLab.htm

Sea anemone

PHYLUM: Cnidaria CLASS: Anthozoa ORDER: Actinaria

For centuries people thought that sea anemones were not animals but a curious form of underwater plant, and even named them for a flower, the anemone. Most sea anemones cannot walk far or swim, and their soft tentacles appear to wave harmlessly in the current like the petals of a flower blown in the wind. However, sea anemones are far from harmless to their prey. Sea anemones are meat eaters that use deadly stinging cells with poisonous barbs to harpoon their victims.

Anatomy and taxonomy

Scientists group all organisms into taxonomic groups based largely on anatomical features. Sea anemones belong to a group of invertebrates called the cnidarians. This group also includes jellyfish, corals, and sea fans.

• **Animals** All animals are multicellular. They get the energy and materials they need to survive by consuming the bodies of other organisms. Unlike plants, fungi, and the members of other kingdoms, animals are able to move around for at least one phase of their lives.

• **Metazoa** Most familiar animal species are metazoans. The exceptions include sponges and colonies of single-celled, animal-like protozoans. Although these organisms can be classed as multicellular, only metazoans have their cells arranged into distinct tissues. A tissue is a body of cells that work together to perform a range of tasks. Nonmetazoans do have cells specialized for certain jobs, but these can function as well alone as in a group. Generally, metazoan tissue cells cannot function alone. The metazoans are divided into several phyla consisting of vertebrates (animals with a backbone) and others. The others are often loosely described as invertebrates, but this is not a taxonomic group. Invertebrates lack bones and

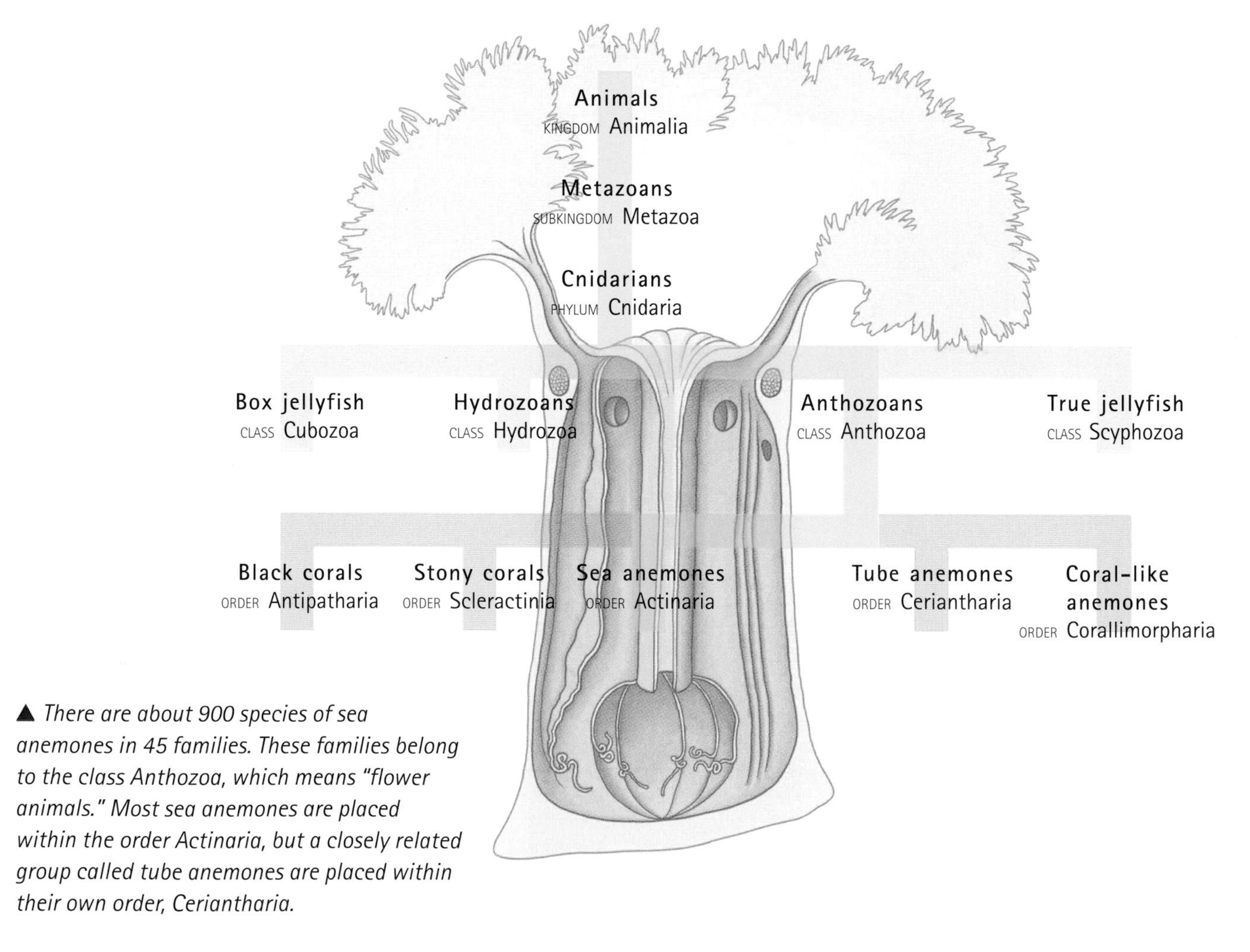

▲ *There are about 900 species of sea anemones in 45 families. These families belong to the class Anthozoa, which means "flower animals." Most sea anemones are placed within the order Actinaria, but a closely related group called tube anemones are placed within their own order, Ceriantharia.*

▶ *Most sea anemones are sessile: they do not move around much but remain attached firmly to rocks or coral, moving only small distances very slowly.*

instead use a range of supportive systems. Most invertebrates (and all vertebrates) have bilateral symmetry: their limbs, skeleton, and muscles are in a mirror-image arrangement around the spine. However, sea anemones and their relatives generally display radial (circular) symmetry as adults.

● **Cnidaria** The phylum Cnidaria includes jellyfish, corals, sea anemones, and box jellies. Cnidarians are believed to have a more primitive body form than any other metazoan animal. Their body is made up of just two tissue layers. Other metazoans have three layers. Cnidarians also have only one body opening. Food passes into the body and waste products and sex cells leave the body along the same route. Many cnidarians adopt two different body forms at different stages in their life. The bell-shape medusa form is free-swimming, whereas the polyp form is tube-shape and generally sessile: it holds fast to a surface. Neither form has a head, and the nervous system is not organized into distinct units. However, cnidarians can possess relatively sophisticated sensory organs, including those that detect and analyze light, orientation, and chemicals in the water column.

● **Anthozoa** This class of cnidarians includes the sea anemones, corals, and sea fans. None of these animals have an independent medusa stage in their life cycle. Instead, they spend their life attached to the seabed as polyps. All anthozoans live in the sea, and many (for example, the corals) are colonial. Over the years, the chalky skeletons made by coral polyps can build up to form huge reefs.

● **Sea anemones** Sea anemones make up the order Actinaria. Unlike corals and other anthozoans, sea anemones never have a calcareous (calcium-containing) skeleton. They do not form colonies but often live in large groups. Sea anemones have many tentacles around the central opening. Of the 45 families of sea anemones, one of the largest is the family Metridiidae, which includes the common plumose anemone. This species lives on hard substrates at a maximum depth of 330 feet (100 m).

FEATURED SYSTEMS

EXTERNAL ANATOMY Sea anemones have radial (round) symmetry. Several tentacles surround the opening to the body cavity. *See pages 1086–1087.*

INTERNAL ANATOMY The body is constructed from two layers of tissues called the epidermis and gastrodermis. Sandwiched between these tissues is a gel-like layer called mesoglea. *See pages 1088–1089.*

NERVOUS SYSTEM Sea anemones do not have a centrally organized nervous system. Instead, the nerves form nets that permeate the body. *See page 1090.*

RESPIRATORY SYSTEM Sea anemones do not have lungs, gills, or any organs for taking in oxygen, nor do they have a circulatory system. Nutrients and respiratory gases are obtained by diffusion across the body's walls. *See page 1091.*

DIGESTIVE AND EXCRETORY SYSTEMS Sea anemones have stinging structures (nematocysts), which are produced in cells in the body wall. The stingers are the largest structures found within any cell in the natural world. *See pages 1092–1093.*

REPRODUCTIVE SYSTEM Sea anemone reproduction is mostly asexual. One technique involves part or all of the body dividing into two to make new individuals. *See pages 1094–1095.*

External anatomy

CONNECTIONS

COMPARE the stinging structures of a sea anemone with those of a ***JELLYFISH***.

COMPARE the polyp body of the sea anemone with the medusa form of a ***JELLYFISH***.

COMPARE the tentacles of a sea anemone with those of an ***OCTOPUS***.

Sea anemones have a tube-shape body that forms a column rising up from the seabed. The top and bottom of the column are called the oral and pedal disks, respectively. The oral disk contains the only opening to the body's internal chamber. This opening is generally called the mouth, but it also has the same function as the anus of less primitive animals. The oral disk is fringed with many tentacles, which range from eight to several hundred in number depending on the size of the animal. At the center of the oral disk is the mouth. The smallest sea anemones have oral disks with a diameter of just 0.25 inch (0.6 cm). The giant sea anemone (which lives around Australian coasts) is the world's largest and can grow to more than 3 feet (91 cm) across. The body is attached to the seabed by the pedal disk. In sea anemones that live on a loose, sandy seabed, the disk grows into a bulblike structure called a physa. The sea anemone drives the physa into the seabed to anchor itself firmly.

Stingers

As with all cnidarians, the tentacles of a sea anemone are lined with stinging cells called cnidocytes. Cnidocytes start out in the body wall as cnidoblasts, and organelles called cnidae, or nematocysts, form within them. As the cnidae mature, the cnidoblasts become cnidocytes. The cnidae fire barbs into anything that touches the tentacle, often delivering poisons, too. A person touching a sea anemone's tentacle might feel the action of the cnidae as merely a sticky sensation. However, smaller animals are quickly overcome by their assault. In some cases, the stings of jellyfish and hydroids can be deadly to humans, too.

▶ **Common plumose sea anemone**
Sea anemones are simple animals. Their body consists of a short column topped by a ring of tentacles surrounding a mouth.

COMPARATIVE ANATOMY

Corals

Like sea anemones, corals exist only in the polyp form. The polyp is cylindrical with tentacles around the mouth, an arrangement similar to that of sea anemones. Their tentacles are less dextrous than a sea anemone's. Cilia on the tentacle's surfaces carry food items to the mouth. Most coral polyps are small, reaching less than 1 inch (2.5 cm) in height, but a few can reach 1 foot (30 cm).

Most corals are colonial: an individual polyp cannot easily be differentiated from its neighbors, and often they share a portion of body wall. Stony corals are reef builders, and unlike sea anemones they build a hard skeleton out of calcium carbonate. When a stony coral polyp dies and decays, the skeleton remains and forms a hard surface for a new generation of polyps to grow on. In that way, a chalky coral reef builds up over thousands of years. Soft corals, sea pens, and sea fans form complicated branched structures. These plantlike structures are made of colonies of many hundreds of polyps.

The entire body is coated with a thick, slimy mucus. The chemicals in this mucus help the sea anemone to recognize other parts of its body, and to identify its relatives. The mucus also acts as a barrier to prevent water from being lost if the animal is exposed to air—for example, when the tide goes out.

Some sea anemones live symbiotically with other sea creatures, especially fish and crabs. Many sea anemones grow on the discarded snail shells occupied by hermit crabs. Both partners appear to benefit from this arrangement. The ordinarily stationary sea anemone is carried along by the crab. The crab gets some protection from its stinging passenger.

Cnidarians have evolved into a huge variety of animals, from sea anemones to reef-building corals and transparent jellyfish. This variety has probably evolved because members of the phylum Cnidaria have two body forms—the polyp and medusa. In general, swimming and floating cnidarians, such as the adult form of the jellyfish, make use of the medusa form. In contrast, sea anemones and other more sedentary cnidarians are all tubular polyps. Despite their many obvious differences, the two forms share the same basic radial structure, with a central mouth surrounded by tentacles.

The mouth *is a slitlike opening in the oral disk.*

tentacles

▲ ORAL DISK AND TENTACLES
Common plumose sea anemone
The radial symmetry can be clearly seen when the sea anemone is viewed from above.

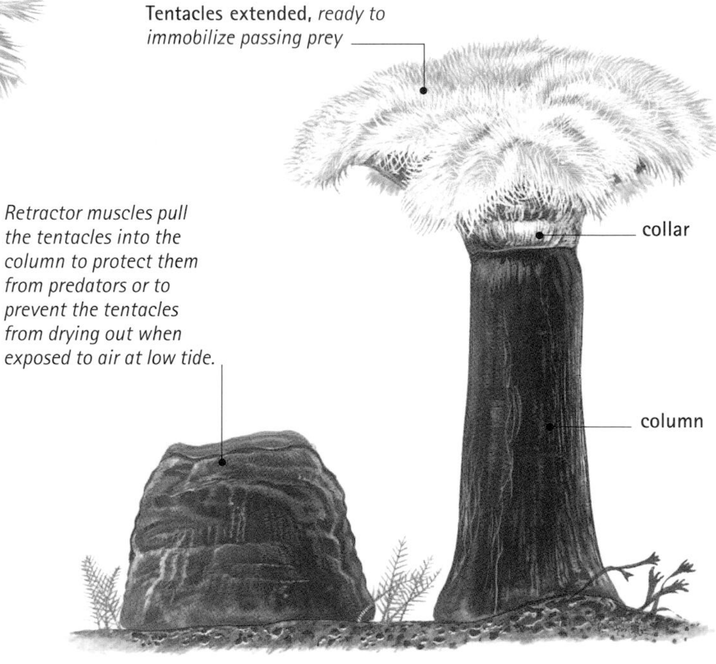

▶ RETRACTED AND EXTENDED TENTACLES
Common plumose sea anemone
When retracted, the tentacles and also the oral disk are covered by the collar.

Internal anatomy

A sea anemone has a single body cavity called the coelenteron, or gastrovascular cavity. As this second name suggests, this cavity is the stomach and circulatory system combined. As is the case for all cnidarians, this cavity has only one entrance: the mouth. The mouth is also the primary exit route.

A sea anemone's mouth forms a slit at the center of the oral disk. The mouth leads to a throatlike pharynx, which connects to the body cavity. The pharynx contains strong circular muscles, which enable it to close tightly. The pharynx also has several grooves called siphonoglyphs. These grooves are lined with ciliated cells. Cilia are tiny hairlike projections on the surface of a cell. When the cilia move in a wavelike pattern they create a current that draws water along the groove and into the sea anemone's body. The siphonoglyphs are one of the few features of sea anemones that are not distributed equally around the pharynx. The area that bears these structures is called the sulcal side. The points without them are asulcal.

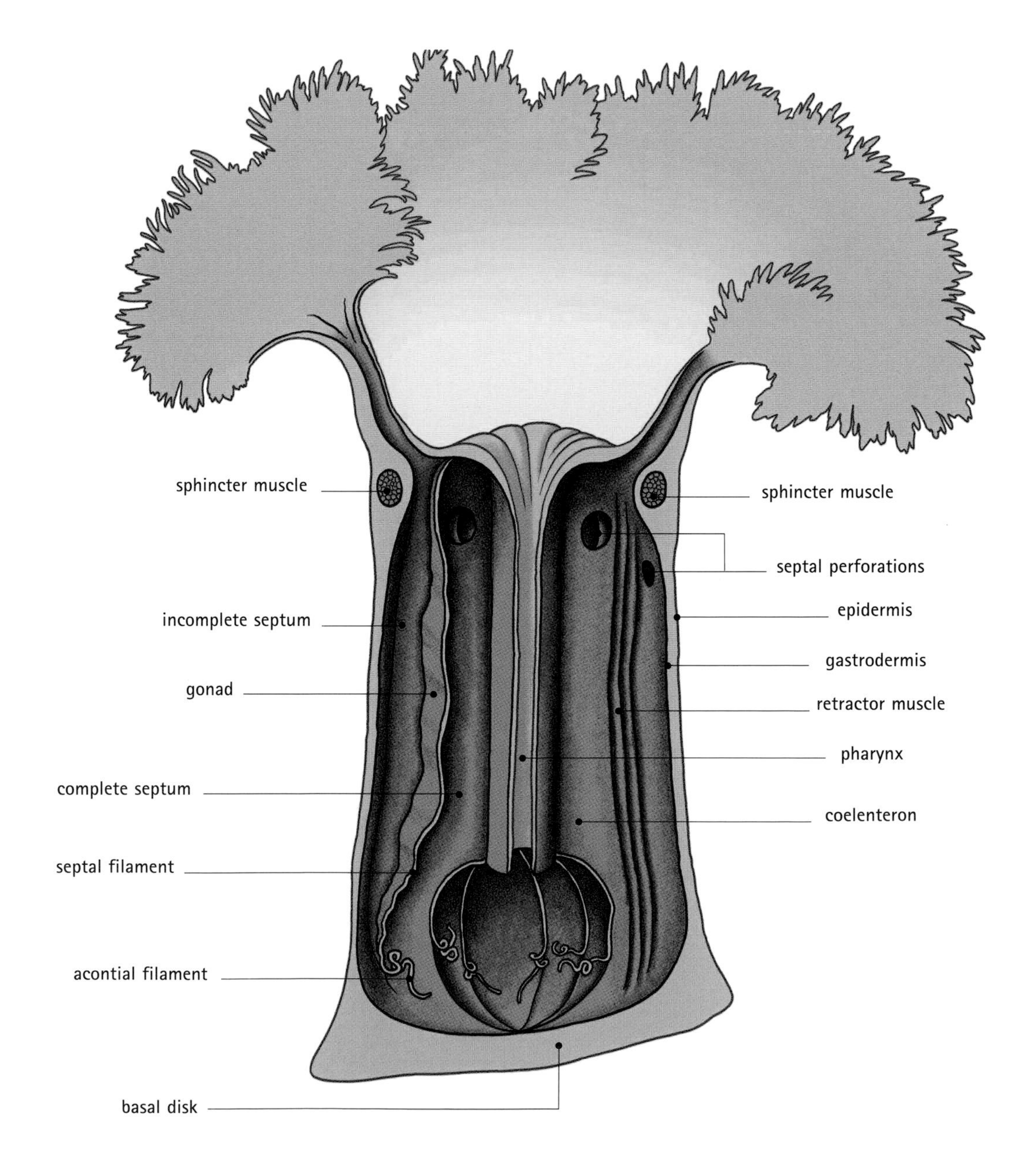

▶ **Common plumose sea anemone**
Sea anemones have a simple internal structure consisting of a cavity called the coelenteron, which is bounded by two thin layers of cells called the epidermis and gastrodermis.

CLOSE-UP

Diploblastic layers

Sea anemones, like all cnidarians, have a body that grows from just two layers of embryonic cells: the ectoderm and endoderm. These develop respectively into the outer layer of cells, the epidermis; and the inner layer of cells, the gastrodermis. This arrangement is described as diploblastic. Other animals are triploblastic. Their body is formed from three embryonic layers: the ectoderm, endoderm, and middle mesoderm. Between the sea anemone's two layers is a gel-like middle layer called the mesoglea. This contains some cells, most of which come from the ectoderm. Unlike the mesoderm of triploblastic animals, the mesogleal cells never develop into tissues in their own right.

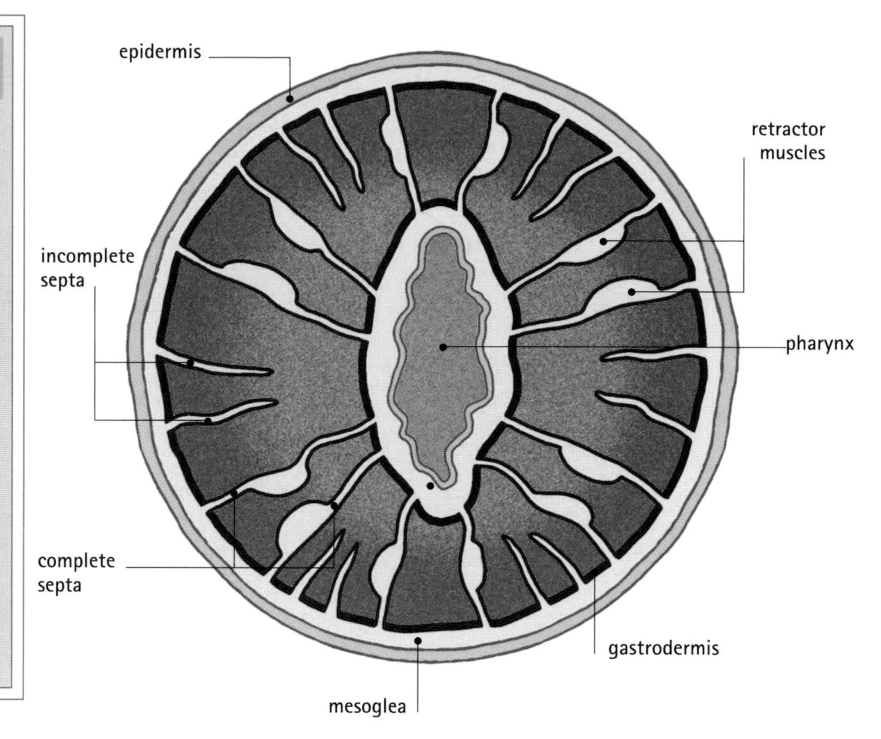

▲ CROSS SECTION *The septa increase the internal surface area of the coelenteron.*

The mouth is also the route by which most contents of the coelenteron leave. However, sea anemones also have pores called cinclides on the side of their body that allow water in the body cavity to be squirted out when a sea anemone contracts its body rapidly. Without them, the increased water pressure inside the coelenteron might cause the sea anemone to burst.

CELL BIOLOGY

Muscle cells

A sea anemone's muscles are relatively simple. Unlike the muscles of other metazoans, cnidarian muscles are not distinct tissues. Instead, they are cells from the epidermis (outer wall) and gastrodermis (inner wall) that have developed several projections at their base that can be contracted. Contraction is produced by fibers of protein inside the projections. The projections of neighboring muscle cells connect to form a contractile sheet between the two tissue layers. Together, the muscle cells can produce circular contractions and longitudinal movements. The circular contractions open and close the mouth and squeeze the body as a whole, and the longitudinal movements shorten the entire body. Cnidarians that live as polyps, such as sea anemones, have more muscle cells in their gastrodermis. Medusoid forms have more muscle cells in the epidermis.

Internal walls

The coelenteron of a sea anemone extends throughout the tubular body and even up inside the tentacles. Internal walls, called mesenteries, extend from the inner surface, or gastroderm. The mesenteries develop in pairs growing out from opposite sides of the gastroderm. Some mesenteries (complete mesenteries) stretch all the way along the body and attach to the pharynx. Incomplete mesenteries are irregular and do not connect with the pharynx. None of the pairs of mesenteries connects in the middle to divide the coelenteron completely.

Mesenteries serve to increase the surface area of the body cavity. This aids absorption of nutrients and gas exchange, which diffuse directly into cells from the surrounding water. The mesenteries also carry many of the sea anemone's muscle cells. The cells are generally part of longitudinal and oblique muscles (the main circular muscles are in the oral disk and pharynx). When the mesenteric muscles contract, they cause the column to shorten. Sea anemones living in the intertidal zone are exposed to air at low tide. By shortening their body until the rising tide covers them with water again, the sea anemones can save water.

Nervous system

Sea anemones do not have a centralized nervous system controlled by a brain. Instead, nerve cells are spread evenly throughout the body, forming two neural nets. One runs between the epidermis and the mesoglea, and the other runs between the gastrodermis and the mesoglea.

The nerve cells of sea anemones and other cnidarians are the most primitive in the animal kingdom. The cells do not have a fatty coating, as do the neurons in many other animals. Also, in these cells, unlike other nerve cells, the impulses can travel in both directions through the cell. Therefore, a stimulus will send out signals in all directions around the body.

Sea anemones appear to have very few sense organs. Tiny hairlike projections on the body surface are probably associated with detecting touch, movement, and chemical changes in the environment. As is the case with other cnidarians, these hairlike projections are most abundant in areas where stinging cells are concentrated. Thus in sea anemones they are most common on the tentacles.

PREDATOR AND PREY

Defense tentacles

Some sea anemones have two kinds of tentacles. Most are feeding tentacles, which usually move together in wavy patterns to trap food; they then transfer food to the mouth. A minority of the tentacles move independently of each other and are called defense tentacles. Defense tentacles are used to search for other sea anemones. Sea anemones do this by extending to about three times their original length and touching the seafloor. Then they retract and repeat the process. If a defense tentacle touches a sea anemone of another species, or one that is not closely related to it, the sea anemone will attack. Then, the tip of the defense tentacle separates and remains stuck to the enemy animal. The presence of the tip causes the surrounding tissues to degenerate, eventually killing the sea anemone that has been attacked.

Puzzling senses

Stinging organelles fire independently of the nervous system. However, biologists think that structures called ciliary cone apparatuses, which develop close to the stingers, may communicate information about the discharged stingers to the rest of the body. It is not clear how this communication occurs, because the cone structures do not appear to be linked directly to the neural nets. Similarly, sea anemones are sensitive to light, but scientists cannot find a light-sensitive receptor. The best guess is that they detect light by nerve cells within or just below the transparent epidermis.

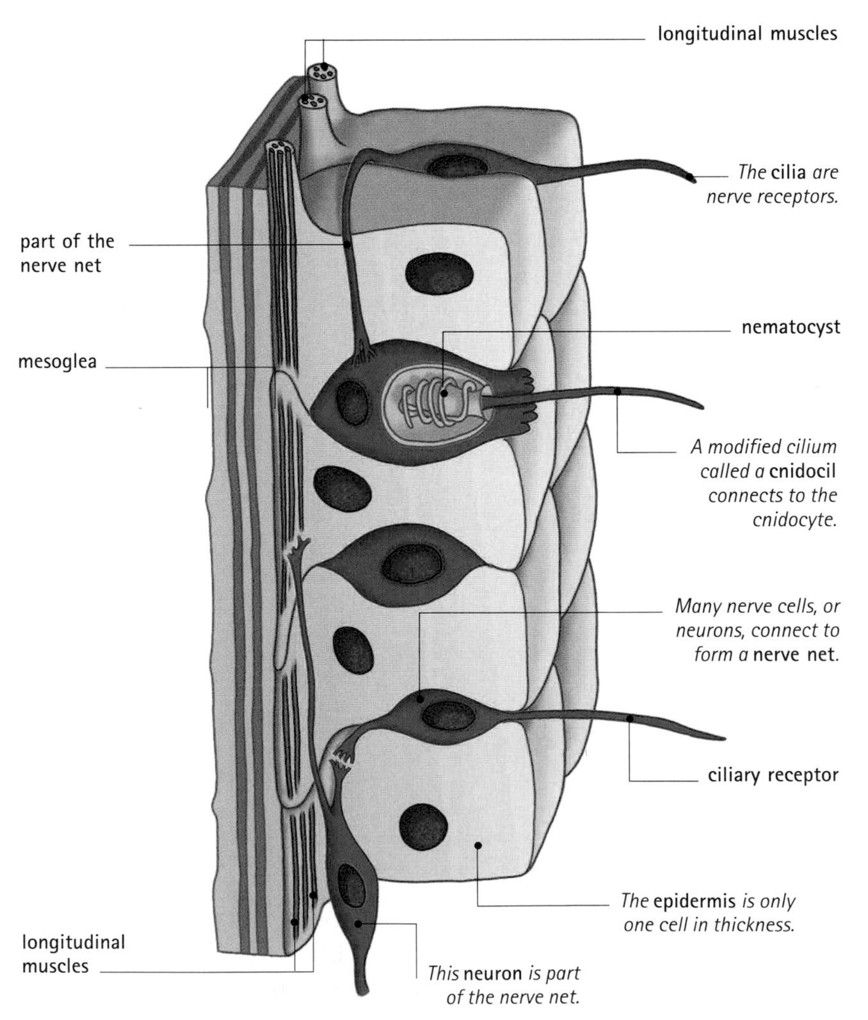

◀ NERVE NET AND CILIA

Hydra

A hydra's nervous system and body wall are similar to those of other cnidarians. A nerve net permeates the body wall and connects with sensory ciliary receptors.

Respiratory system

Sea anemones do not have a blood supply or any other circulatory system to distribute nutrients around the body. Everything the animal needs, however, is contained in the coelenteron and the surrounding water. Sea anemones rely on the process of diffusion for nutrients to get where they need to go. Diffusion is a slow process, and it is inefficient at distributing substances around the whole internal volume of large sea anemones. The problem is partly solved by the mesenteries, which form side chambers, or septa (singular, septum), in the coelenteron. This arrangement increases the surface area across which substances can diffuse into the body.

Nutrients from digested food are absorbed directly into the cells that need them. In the same way, oxygen dissolved in seawater diffuses straight into the cells. Gas exchange takes place through both the external and the internal body walls. The layers of cells that make up the epidermis and gastrodermis are very thin, and thus the distances across which gases and nutrients must diffuse is kept to a minimum. It is the gel-like mesoglea that gives a sea anemone's body thickness.

Anaerobic respiration

Some species of sea anemones bury themselves in soft sediments, such as sand or silt. Often just their oral disk and tentacles extend above the substrate. The remainder of the body is anchored in the ground by the physa. Some species bury themselves completely in sand when disturbed. They do this by folding in their tentacles and swelling the physa so it is wider than the rest of the body. Then the retractor muscles in the mesenteries pull down, and the oral disk slips under the sand. With only a small amount of the body in contact with free-moving water, these burrowing sea anemones have difficulty getting enough oxygen. They are forced to respire (release energy from their food) anaerobically—that is, without oxygen.

CONNECTIONS

COMPARE the circulatory function of the coelenteron with the open circulatory system of an ***ANT*** or another insect, and with the closed system of a ***HUMAN*** or another mammal.

COMPARE the way a sea anemone takes oxygen through its surfaces with the way a ***BULLFROG*** takes oxygen through its skin.

◀ *This burrowing sea anemone lives on the Great Barrier Reef, Australia, and grows to about 16 inches (40 cm) across. Its column lies below the surface of the sand, and the tentacles can be withdrawn quickly under the sand to avoid predators.*

Digestive and excretory systems

Sea anemones are meat eaters. They eat a variety of other aquatic animals, especially small fish and crustaceans. Sea anemones are able to subdue more lively prey with their stinging organelles, or nematocysts. These features are unique to cnidarians. No other group of animals has anything like them, although some sea slugs that prey on sea anemones have found a way of recycling their victims' weapons and arranging them on their own back. These sea slugs use the nematocysts as a defense against predators.

A single stinging structure is called a nematocyst. It is supported by several surrounding cells and the underlying mesoglea. Each nematocyst contains a stinger, or cnida. Scientists believe that cnidae are the largest and most complex intracellular objects (that is, objects located within a cell) in the natural world. Scientists do not understand how cnidae form inside the cnidocyte, although they are probably secreted by the Golgi apparatus. The Golgi apparatus is the cellular system that organizes material to be removed from the cell.

IN FOCUS

Zooxanthellae

Like corals and other cnidarians, sea anemones have a close relationship with single-cell algae called zooxanthellae. The algae live mostly in the epidermis and gastrodermis of the sea anemone, although a few are found in the mesoglea. It is these algae that give many sea anemones their bright colors. The algae photosynthesize inside their host, producing sugars. As with corals, sea anemones that live in warm, shallow waters bathed in plenty of sunlight receive a large proportion of their nutrients from the algae. In return, the sea anemones provide a safe place for the zooxanthellae to live. A relationship of this sort, in which both organisms benefit, is called symbiotic.

Protein capsules

Cnidae are capsules made of protein. A long tube is coiled at the bottom of the capsule. When a cnida is stimulated, the contents of the capsule turn inside out, bursting out through a flap at the top of the cell. As a result, the tube straightens and fires rapidly from the cell. The tube is often equipped with barbs or spikes

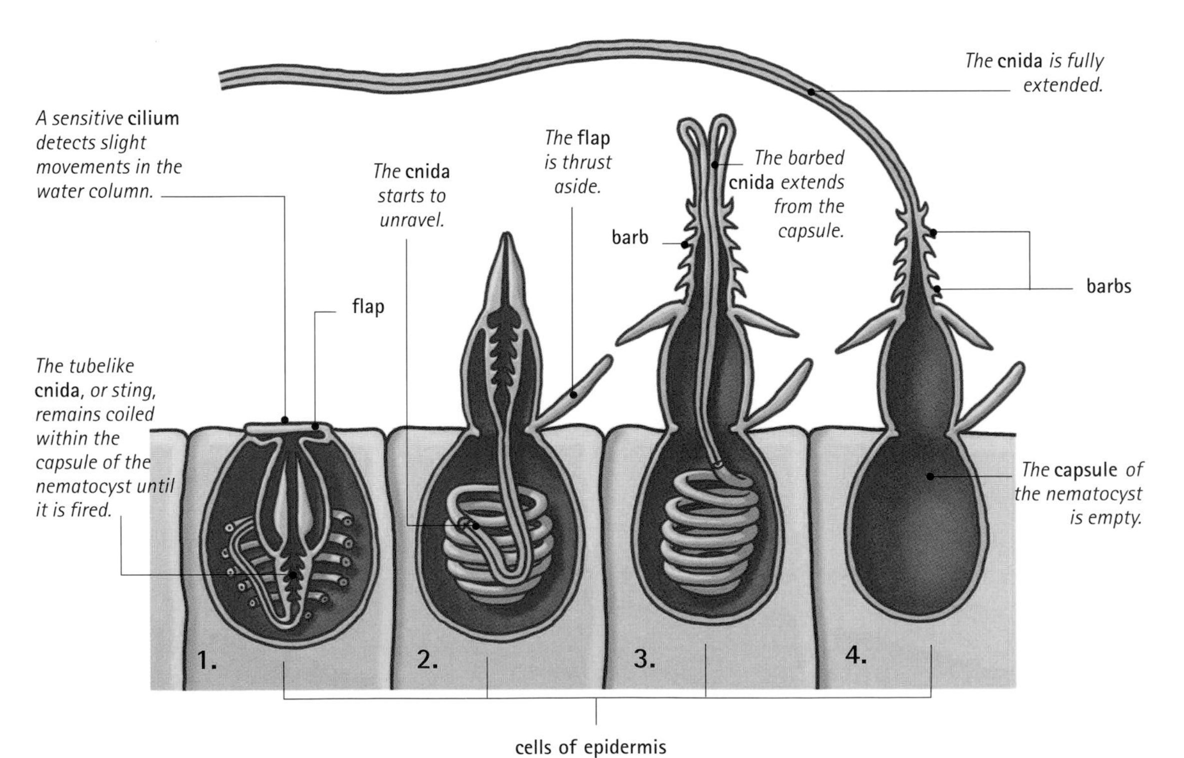

◀ **NEMATOCYST AT REST AND FIRING**
Most of the time, the cnida, or stinger, remains coiled within the nematocyst capsule in a cell of the epidermis (1). When a cilium is stimulated, the cnida bursts out through the flap and turns inside out (2 and 3). If the cnida penetrates prey, barbs may deliver toxins that subdue it.

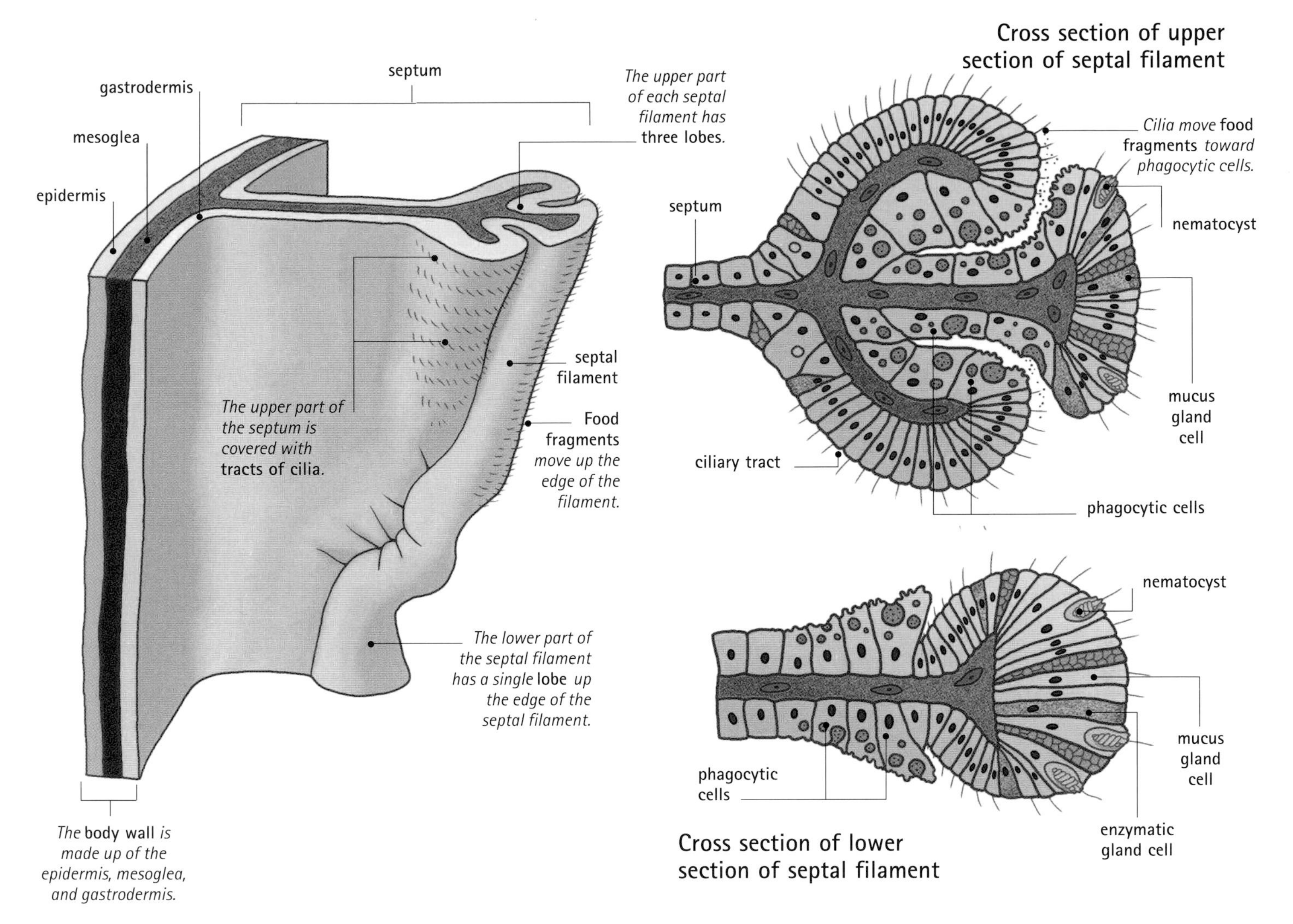

that stab the prey. Most barbs also deliver a cocktail of toxins. The sting is fired after a cilium on the surface of the nematocyst detects changes in pressure created by water currents. The sea anemone's weapons are probably tuned to certain types of electrical currents that are produced by its preferred prey. Scientists cannot agree on the mechanism by which the cnida is fired, but it is triggered by an influx of water, by tension built into the system during its development, or by squeezing the capsule with microtubules.

Feeding behavior

Most sea anemones use their tentacles to capture prey and transport them to the mouth. They use their nematocysts to subdue prey. However, one group of large sea anemones does not use their tentacles to do this. Instead, they extend their pharynx upward to snap up prey. Prey may still be alive even after having been swallowed. Threads, or acontia, equipped with many nematocysts extend down from the pharynx. The threads continue to sting any prey that have been swallowed alive. Some sea anemones also fire these threads out of their mouth or pores to aid in capturing prey.

Cells lining the mesenteries produce enzymes that digest food. Other cells have cilia, and the currents the cilia produce help mix the digestive juices. This first stage of digestion breaks food into large molecules, such as starches and polypeptides. These are engulfed by cells lining the cavity wall, and the final stages of digestion take place inside food vacuoles. The waste materials from this process are released by the cell back into the body cavity. Waste products build up inside the sea anemone, so periodically it flushes the contents of its coelenteron out through the pores. The sea anemone flushes its coelenteron by rapidly contracting its body column.

▲ SEPTUM
The septum consists of two layers of gastrodermis surrounding a layer of mesoglea. The edge of the septum is convoluted, and the upper portion is covered with cilia that help to circulate water through the coelenteron.

Reproductive system

Sea anemones reproduce mainly by asexual means, although most species can also reproduce sexually. Some sea anemones are hermaphrodites, producing both male and female sex cells. Other species, however, are gonochoristic; that is, the individuals are differentiated into two distinct sexes.

Asexual reproduction

There are two forms of asexual reproduction practiced by sea anemones: longitudinal fission and pedal laceration. Longitudinal fission quickly produces large groups of clones, or genetically identical individuals. In this process one sea anemone divides to make two individuals. This is a complex event and is poorly understood.

Longitudinal fission begins with the sea anemone's tubular body stretching sideways. This action stimulates areas in the middle region of the body to begin to degenerate, with the cells in the two tissue layers dying. The body becomes thinner in the area where the cells die, making it easier for fission to occur. The two new individuals may edge away from each other to give themselves enough space (sea anemones can move short distances when necessary). Longitudinal fission takes several days to complete, but it can produce a large group of clones in a matter of months if the conditions are right.

COMPARATIVE ANATOMY

Copulation

Only one species of sea anemone is known to copulate, although there are likely to be others. *Sagartia troglodytes*, which lives in coastal waters of the eastern North Atlantic, is gonochoristic. The female moves beside a male to begin copulation. Their pedal disks are pressed together to create a chamber for the two sets of sex cells. This position is held for several days as the fertilized eggs inside develop into planulae. The larvae are then released. Copulatory behavior may be a response to powerful water currents, which could wash away sex cells released into the open water before they have a chance to mix. Coupling ensures that the sex cells have ample opportunity for mixing.

▼ *A planula larva of a burrowing sea anemone develops into a more complex larva, which, like the adult sea anemone, has a ring of tentacles around the mouth.*

The second type of asexual reproduction, pedal laceration, also produces clones, but it does so in a slightly less dramatic way. In this process the pedal disk grows away from the body. As the disk spreads, the rest of the sea anemone's body travels with it, leaving fragments of the disk behind. These fragments develop into new individuals.

Sexual reproduction

A sea anemone's gonads develop from the gastrodermal cells of its mesenteries. The gonads release their sex cells—sperm and eggs—into the coelenteron. These cells are then ejected from the body into open water, where they mix with the cells released by neighboring sea anemones. The fertilized egg, or zygote, develops into a planula larva. This is a solid tube of cells made up of a central core of internal

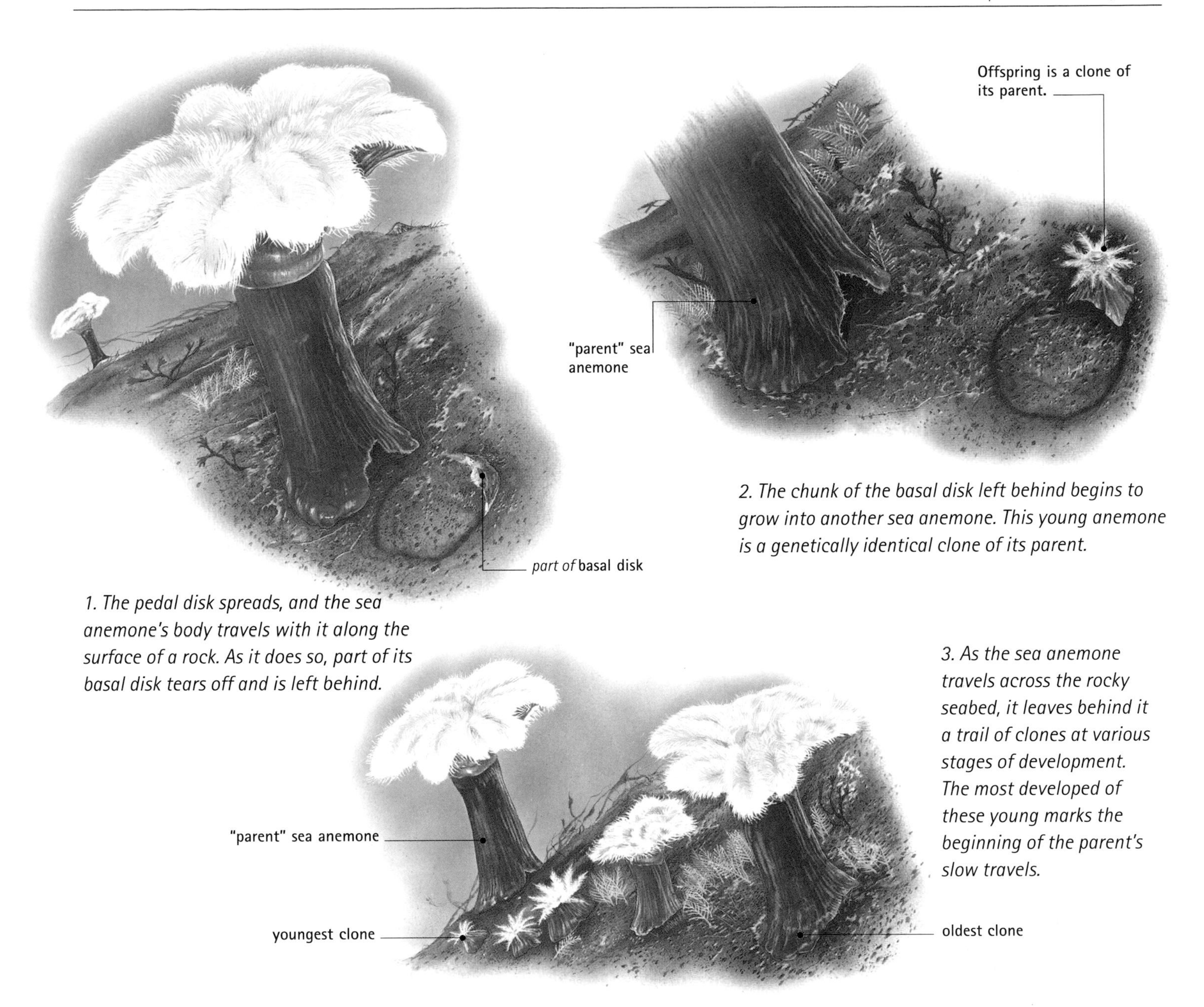

1. The pedal disk spreads, and the sea anemone's body travels with it along the surface of a rock. As it does so, part of its basal disk tears off and is left behind.

2. The chunk of the basal disk left behind begins to grow into another sea anemone. This young anemone is a genetically identical clone of its parent.

3. As the sea anemone travels across the rocky seabed, it leaves behind it a trail of clones at various stages of development. The most developed of these young marks the beginning of the parent's slow travels.

cells called endoderm and a surface of ciliated external cells called ectoderm. The larva has a brief planktonic phase, during which it feeds on plankton. The larvae of all anthozoans do the same. The planktonic phase allows growing anemones to disperse before settling on the seabed and growing into the adult polyp phase.

Sea anemones have polypoid bodies and have lost (or never had) a medusoid phase. Most other cnidarians exist as both forms, for however briefly, producing planulae while in the medusoid form. This fact and several other anatomical similarities between medusae and sea anemones have led some biologists to argue that sea anemones are not polyps at all, but sedentary medusae.

▲ PEDAL LACERATION
Common plumose sea anemone
Pedal laceration is a form of asexual reproduction in which a sea anemone produces young that are genetically identical to itself; that is, they are clones.

Nurturing young

In some hermaphrodite species of sea anemones, fertilization takes place inside the sea anemone's body. The young are retained for the first part of their development before being released. Release is generally through the mouth, but in at least one species, the northeast Pacific sea anemone, the larvae are released through a pore in the tip of each tentacle.

TOM JACKSON

FURTHER READING AND RESEARCH

Anderson, D. T. (ed.) 2001. *Invertebrate Zoology*. Oxford University Press: New York.

Shick, J. Malcolm. 1991. *A Functional Biology of Sea Anemones*. Chapman and Hall: New York.

Sea horse

CLASS: Osteichthyes ORDER: Syngnathiformes
FAMILY: Syngnathidae SUBFAMILY: Hippocampinae

The 35 or so species of sea horses live among seaweeds, sea grasses, and corals. The Indo-Pacific spotted sea horse, for example, lives near coral reefs in shallow warm waters in parts of the Indian and Pacific oceans. There, it feeds on small zooplankton (animal plankton) such as copepods and the larvae of shrimp, crabs, and fish.

Anatomy and taxonomy

Scientists categorize all organisms into taxonomic groups based partly on anatomical features. Together with about 200 species of pipefish, sea horses make up a group of highly unusual-looking fish in the family Syngnathidae. Scientists disagree on how to classify sea horses and pipefish, and the scheme used here is only one of several in common use. Like other sea horses, the Indo-Pacific spotted sea horse is vulnerable because it has been collected as a souvenir for tourists and for use in Chinese medicine, and some of its favored habitats are threatened by destruction or alteration.

- **Animals** Sea horses, like other animals, are multicellular. Animals eat other organisms, breaking them down into simple molecules, which are used to provide energy for the animal or to build new tissues. Animals differ from other multicellular life-forms in their ability to move from one place to another (in most cases, using muscles). They generally react rapidly to touch, light, and other stimuli.

▶ *Sea horses and pipefish are unusual-looking fish that make up the family Syngnathidae. The subfamily Hippocampinae comprises sea horses alone. Only living groups are shown in this family tree.*

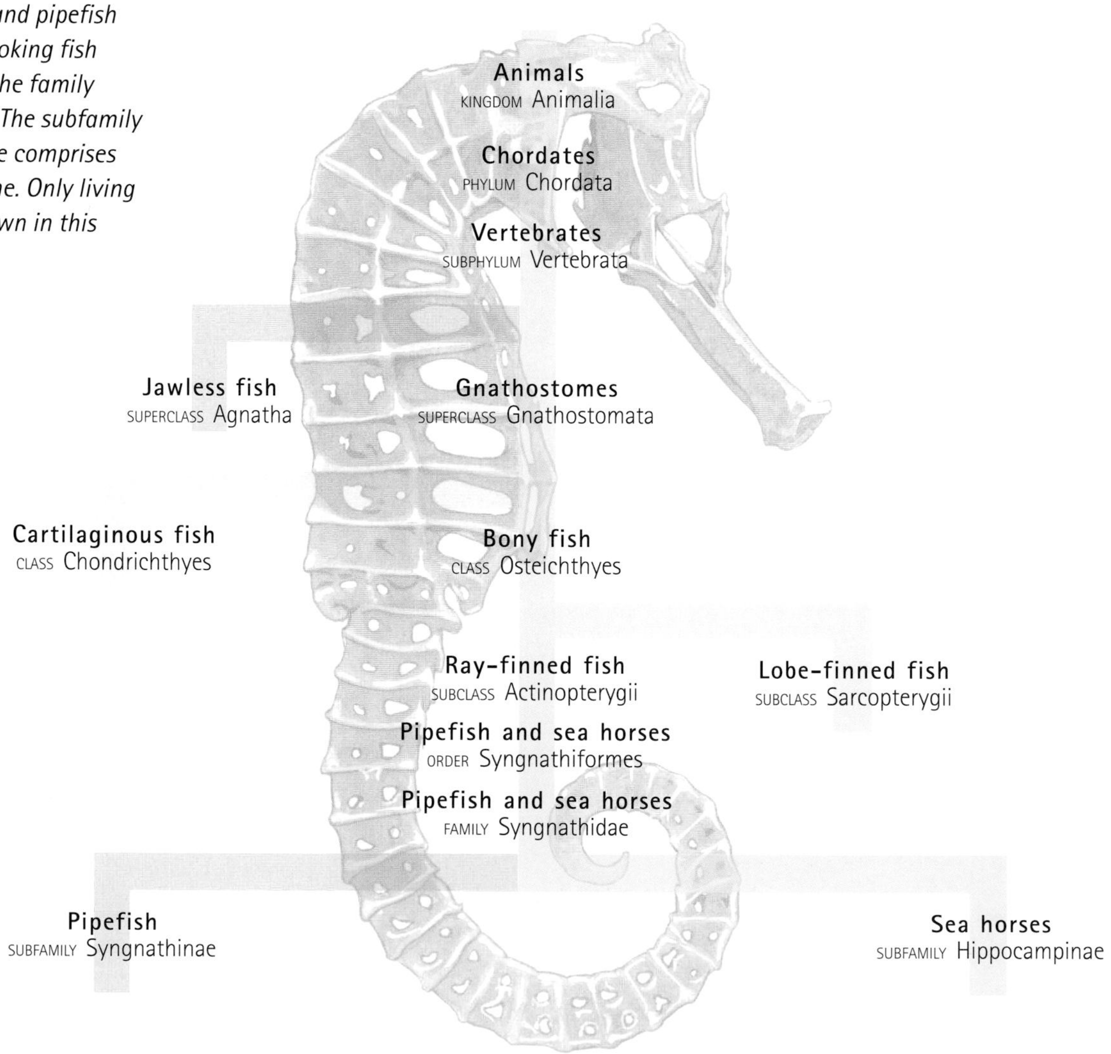

Animals
KINGDOM Animalia

Chordates
PHYLUM Chordata

Vertebrates
SUBPHYLUM Vertebrata

Jawless fish
SUPERCLASS Agnatha

Gnathostomes
SUPERCLASS Gnathostomata

Cartilaginous fish
CLASS Chondrichthyes

Bony fish
CLASS Osteichthyes

Ray-finned fish
SUBCLASS Actinopterygii

Lobe-finned fish
SUBCLASS Sarcopterygii

Pipefish and sea horses
ORDER Syngnathiformes

Pipefish and sea horses
FAMILY Syngnathidae

Pipefish
SUBFAMILY Syngnathinae

Sea horses
SUBFAMILY Hippocampinae

▶ *The spotted sea horse varies in color from pale yellow or cream, speckled with small black spots and blotches, to almost entirely black. Adult spotted sea horses can grow up to 12 inches (30 cm) long, from the tip of the prehensile (gripping) tail to the bony projections forming a coronet on the top of the head.*

• **Chordates** At some time in its life cycle, a chordate has a stiff, dorsal (back) supporting rod called the notochord that runs all or most of the length of the body.

• **Vertebrates** In living vertebrate animals, the notochord develops into a backbone (spine, or vertebral column) made up of units called vertebrae. The vertebrate muscular system that moves the body consists primarily of muscles in a mirror-image arrangement on either side of the backbone or the notochord; in this way, vertebrates have bilateral symmetry about the skeletal axis.

• **Gnathostomes** Gnathostomes are jawed fish, as opposed to hagfish and lampreys (agnathans), which lack proper jaws. To breathe, gnathostomes have gills that open to the outside through slits. Jawed fish also have fins that include those arranged in pairs, such as the pectoral (shoulder) fins.

FEATURED SYSTEMS

EXTERNAL ANATOMY Sea horses have a horselike head with projections at the top, called a coronet, and a long snout. The body is held upright and is covered with bony plates. The fins are small, and the tail is usually coiled. *See pages 1099–1100.*

SKELETAL AND MUSCULAR SYSTEMS The bony plates act as an outer skeleton. The arrangement of muscles and bony plates enables a sea horse's tail to grasp objects. *See pages 1101–1102.*

NERVOUS SYSTEM The sea horse's most important senses are sight and touch. The eyes detect colors and can focus on small prey. Both vision and touch are crucial in greeting and courtship rituals between male and female sea horses. *See page 1103.*

CIRCULATORY AND RESPIRATORY SYSTEMS Sea horses have unusual grapelike gills. They contain bulbous tubes of blood-rich tissues through which seawater flows. *See pages 1104–1105.*

DIGESTIVE AND EXCRETORY SYSTEMS The sea horse's gut is simple and typical of fish that mostly eat a continuous diet of small animals. *See pages 1106–1107.*

REPRODUCTIVE SYSTEM Sea horses are some of a small number of fish in which the male becomes "pregnant." The female sea horse lays her eggs in the male's brood pouch, which supplies the developing young with oxygen and nutrients and removes their wastes. *See pages 1108–1109.*

• **Bony fish** Sea horses belong to the class Osteichthyes (bony fish), the major group that includes more than 95 percent of all fish. Bony fish, as the term implies, have a skeleton of bone, in contrast to members of the class Chondrichthyes—cartilaginous fish such as sharks, skates, rays, and chimeras—that have a skeleton made of cartilage.

• **Ray-finned fish** Almost all bony fish, including sea horses and pipefish, belong to the subclass Actinopterygii (ray-finned fish). The principal feature that distinguishes ray-finned fish from the eight species of the subclass Sarcopterygii (fleshy-finned fish) is the presence of bony rays that support thin, flexible fins.

◀ *The thorny sea horse lives in the Indian and Pacific oceans and grows up to 6.7 inches (17 cm) long. It has a long snout, sharp spines, and a five-pointed coronet.*

• **Pipefish and sea horses** Pipefish and sea horses belong to the order Syngnathiformes. The name is derived from the Greek words *syn,* meaning "together," or "fused," and *gnathus,* meaning "jaws," and refers to the tubelike mouth of these fish. Unlike many other fish, pipefish and sea horses do not have a body that is streamlined and suited to life in open water. Instead, these fish swim slowly and rely on camouflage and body armor to avoid predators and protect themselves against attack. The body is long and thin and encased in armorlike rings and plates of bone. Over millions of years of evolution, these fish have dispensed with pelvic fins because these are no longer required as stabilizers for forward swimming. Sea horses and pipefish feed on small animals that float in the water and have no need to swim quickly. All sea horses and some pipefish are unusual in that the female lays eggs but passes them to her male partner. The male incubates the eggs in or on his body until they hatch. Most species of sea horses and pipefish live in the sea, but some inhabit brackish water (diluted seawater), and a few occupy freshwater habitats.

• **Pipefish** Most pipefish species (subfamily Syngnathinae) have a nearly straight, long body. However, a few members of the subfamily—the bizarre-looking sea dragons—have an appearance between that of pipefish and sea horses. Strange leaflike appendages extend from the body of sea dragons, breaking up their outline and giving them effective camouflage among brown seaweeds. Most pipefish propel themselves with their tail, unlike sea horses.

• **Sea horses** These fish, unlike pipefish (other than sea dragons), have a head that is positioned more or less at a right angle to the body. In sea horses, the tail lacks a fin and is prehensile, being used for grasping strands of seaweed or sea grass rather than for swimming. Sea horses usually swim upright in the water, propelling themselves forward using the dorsal (back) fin and pectoral fins. The orientation of the head and neck, and the presence of spines that look like a mane, gives sea horses a superficially horselike appearance, for which they are named.

IN FOCUS

The number of sea horse species

A group of North American scientists carried out a review of sea horse classification in 1999 and identified some 35 species. Since then, scientists have discovered another species, the pygmy sea horse, in the western Pacific Ocean, where it lives among branched coral. The pygmy sea horse is tiny, just 0.7 inch (1.6 cm) long. The discovery suggests that there may be more sea horse species yet to be identified in the oceans. In addition, as scientists carry out further analyses of the structure, genetics, and biochemistry of existing sea horse specimens, it is probable that some species may be split into two or more species of sea horses.

External anatomy

The spotted sea horse barely looks like a fish at all, except for its fins. Its body, unlike that of most other fish, is not streamlined in a way that makes for easy movement through water. Instead, the sea horse is shaped for camouflage and armored for protection. It is difficult to see a sea horse floating upright among seaweed, sea grass, or coral.

The spotted sea horse generally varies in color from pale yellow or cream speckled with small black spots and blotches to almost entirely black. Like other sea horses, it has the astonishing ability to change color, either to match the shade of its background or to make a bright display to ward off an attacker or attract a mate. Spiny projections made of bone, such as the coronet on top of the head and spines along the back, act as a deterrent to animals that may try to eat sea horses. The projections also help break up the fish's outline and aid camouflage. A large coronet can be a sign of a healthy, successful male, particularly one likely to attract a mate.

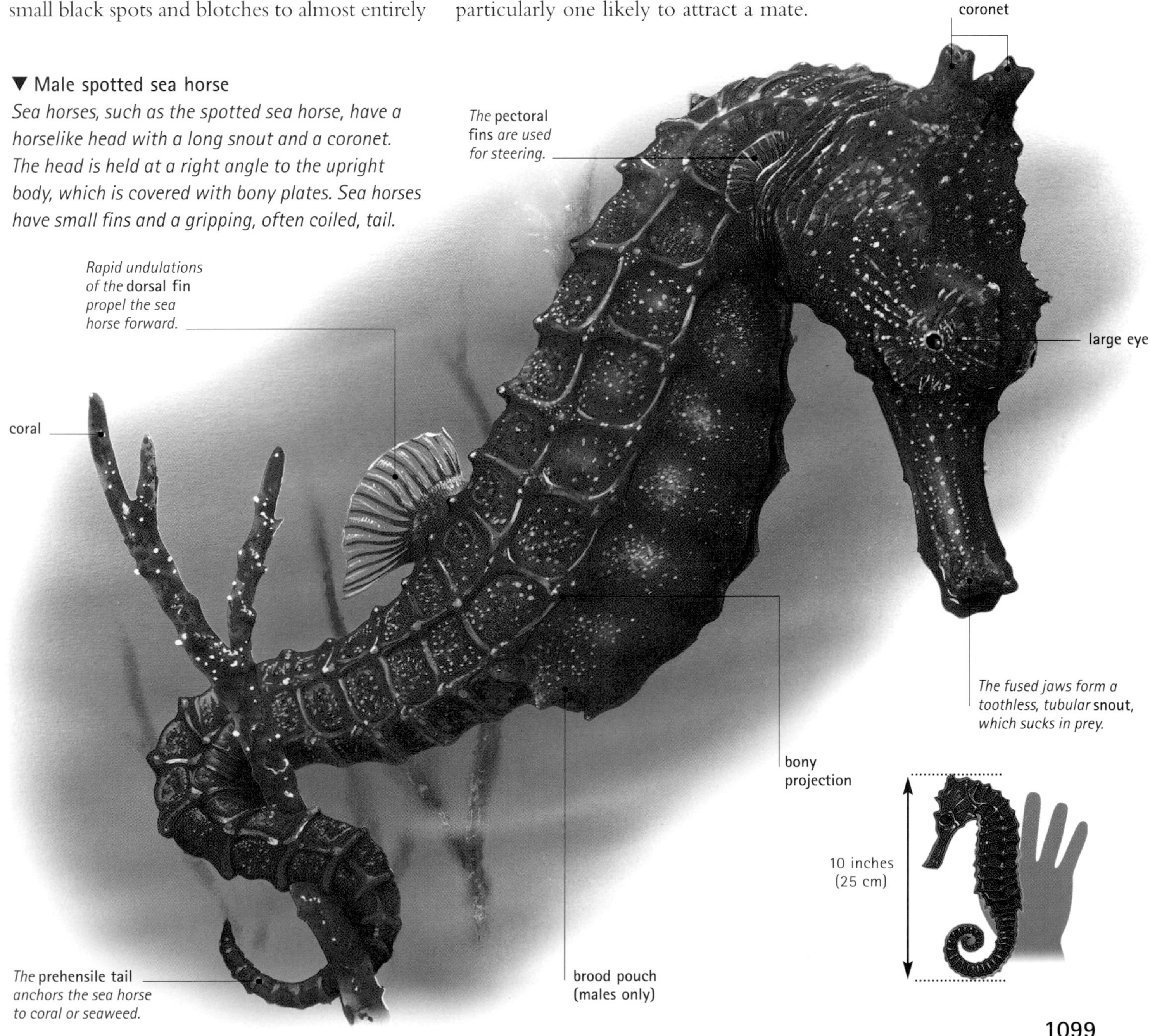

▼ Male spotted sea horse
Sea horses, such as the spotted sea horse, have a horselike head with a long snout and a coronet. The head is held at a right angle to the upright body, which is covered with bony plates. Sea horses have small fins and a gripping, often coiled, tail.

EVOLUTION

Losing fins

Sea horses and pipefish are closely related to fish called sticklebacks. They have in common bony plates and a small mouth. The sequence of body forms—stickleback, tube-snout, pipefish, and sea horse—suggests how ancestral sea horses might have evolved from a more conventionally shaped fish with a full complement of fins for swimming. Long, upright forms have evolved over millions of years to become strangely shaped fish that do not have pelvic fins and, in the case of sea horses, lack the tail (caudal) fin as well.

Scientists often use the shape and number of bony projections, along with the precise arrangements of plates and bony rings in the exoskeleton, to distinguish one species of sea horse from another. In addition, some species of sea horses grow skin tendrils that look like curly hairs. The tendrils become encrusted with algae and microorganisms, and these improve the sea horse's camouflage.

Eyes and vision

The sea horse's two eyes are positioned at the sides of the head, giving the fish a wide field of vision for detecting both prey and potential attackers. The eyes can move independently of each other, enabling the sea horse to observe a large part of its environment without having to move the rest of its body. The eyes are angled slightly together at the front, so their fields of vision overlap. This overlap gives the sea horse binocular vision in front of the head, which allows the fish to judge distances and home in on prey such as small crustaceans. The sea horse uses its tubelike snout like a sucking pipette to draw prey into the mouth. Eyesight is of great importance to the sea horse, and anything that clouds the water puts the animal at a disadvantage in seeking prey.

The gill openings, one at either side of the head, are very small and look rather like ear openings. They are the exits from gill chambers that contain highly unusual-looking gills, which resemble bunches of grapes, rather than the leaf-shape gills of most other fish.

▶ *Sea dragons are a type of pipefish, which are close relatives of sea horses. The elaborate leafy appendages of a sea dragon keep it well camouflaged among brown seaweeds.*

Skeletal and muscular systems

Sea horses move slowly, picking their way amid the underwater vegetation or coral where they hide themselves. The shape of the body, and the way in which the skeletal and muscular systems work, is suitable for slow, delicate movements rather than fast swimming as in many other fish.

Armorlike outer skeleton

The sea horse's outer skeleton, or exoskeleton, is a bony suit of armor that makes the fish unpalatable. The outer skeleton is made up of bony plates and ridges that grow just beneath the skin. The trunk is encased in seven rows of bony plates that run from the neck to the base of the tail. The tail is protected by four rows of plates. The plates overlap to allow movement; but in the sea horse, unlike most other kinds of fish, the body and tail can bend only slightly from side to side. Sculling movements of the pectoral fins and undulatory waves passing along the dorsal fin propel the sea horse forward. Flicked back and forth or tilted, the pectoral fins also adjust the sea horse's position in the water or steer.

The sea horse uses its tail to grasp strands of seaweed, rocks, or coral. This action anchors the sea horse to the bottom and keeps it from being swept away by water currents. In swimming, the tail is either extended or coiled up toward the belly. The tail is also used for touching in greeting and in courtship rituals, when two sea horses meet and entwine their tails. During mating, the tails anchor the partners together. Rival male sea horses use their tails in bouts of wrestling when fighting over a female. Adult males can be distinguished easily from females by their brood pouch, which is located in the lower abdomen. The pouch bulges when the male is incubating eggs.

CONNECTIONS

COMPARE the swimming action of a ***TROUT*** with that of a sea horse. The trout swishes its tail from side to side to propel itself forward, whereas the sea horse propels itself mainly by rapid undulations of the dorsal fin.

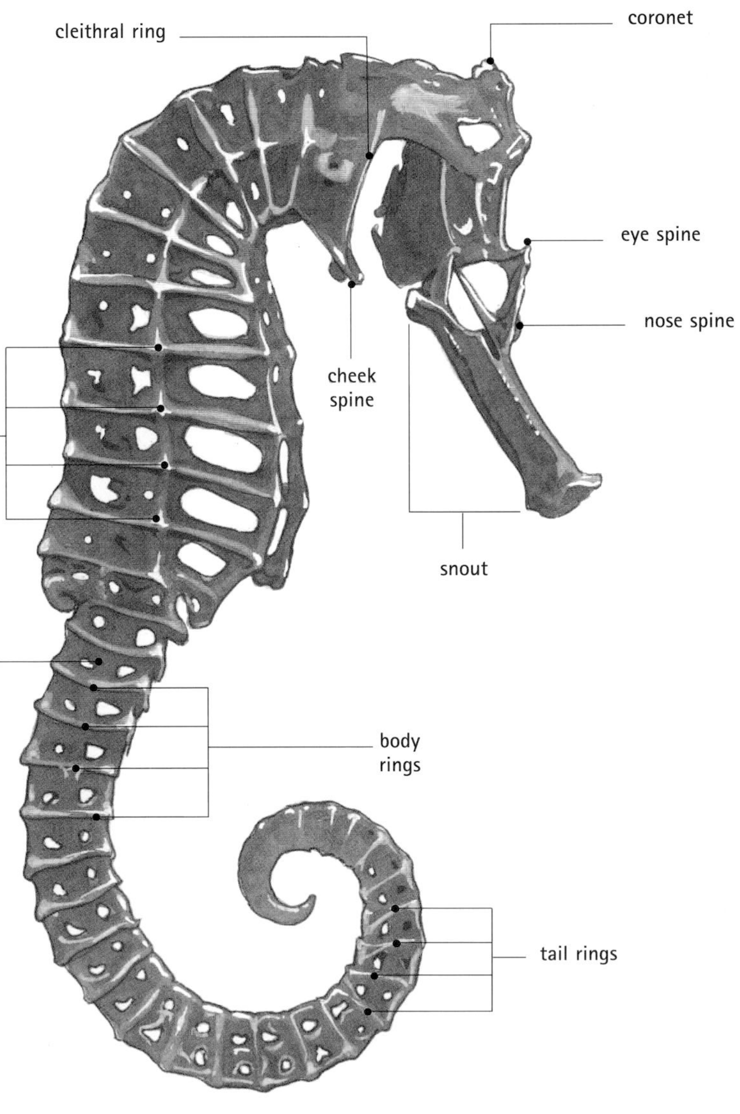

cleithral ring
coronet
eye spine
nose spine
cheek spine
studs
snout
body rings
tail rings

The body is encased in an outer skeleton of bony plates. *Each plate has a raised area, which forms a ring around the body.*

▶ SPINE AND SKULL
Spotted sea horse
As with other bony fish, the sea horse skeleton comprises jaws, a skull, and a backbone. Sea horses do not have scales like most other fish. Instead, their whole body is covered by rings of bony, studded plates beneath the skin that protect the fish and make them unpalatable to would-be predators.

Skeleton and muscles

Within the body, the sea horse has a skeleton, which (as in other bony fish) comprises a skull, jaws, and a backbone. The parts of the skeleton are moved by muscles that pull on the bones, which act as levers. Skeletal muscles contract to produce power and movement. Muscles cannot extend of their own accord, so they are usually arranged in antagonistic sets, one set working against the other. While one set is contracting, the other is relaxing and lengthening, ready to contract again.

Most fish swim by side-to-side movements of the tail. Sea horses swim differently. Their tail is most flexible when it moves vertically, and it is used for grasping weeds and coral rather than swimming. In sea horses, antagonistic sets of muscle blocks, called myomeres, are arranged to bend the spine in a vertical direction rather than from side to side. These muscle blocks connect between the backbone and the bony plates beneath the skin. When the muscle blocks below the backbone contract, they bend the spine downward, and the overlapping bony plates move one against the other, bending at their joints. Tail flexing and coiling in the ventral direction can be rapid and powerful, allowing the sea horse to grasp objects for anchorage or push itself off the seabed.

Dorsal and pectoral fins

Sea horses swim forward or upward. Forward swimming is accomplished largely using the dorsal fin. A wavelike undulation moves quickly down the fin from top to bottom, propelling the fish upward and forward. When the fish is swimming more rapidly, its body tilts forward at about 45 degrees and the propulsive force is more directly forward, with little vertical lift. The two pectoral fins steer the body. The sea horse's anal fin is small and, other than acting as a rudder to help prevent rolling, is not involved in most forms of swimming. However, by beating its anal fin from side to side, the sea horse can swim very slowly upward or help maintain its vertical position in the water.

▶ *The golden sea horse uses its tail for grasping a strand of seaweed. This action anchors the sea horse to the seabed and prevents it from being swept away by water currents.*

Like the body, the tail is encased in **bony plates**. *White's sea horse has 11 bony plates on the trunk and between 33 and 36 on the tail.*

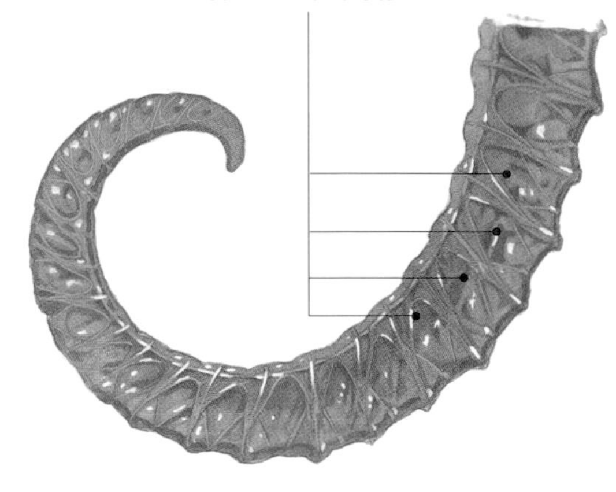

▲ BONY PLATES IN TAIL

White's sea horse

Rows of plates protect the sea horse's tail. The rings limit the fish's maneuverability, but since they are narrower on the tail, that part of the body bends more easily. The plates overlap to allow the tail to bend forward and backward but only slightly from side to side.

Nervous system

The nervous system of sea horses is based on the same plan as that of bony fish such as trout and sailfish. The central nervous system (CNS) consists of a brain and spinal cord. The CNS is connected via nerves of the peripheral nervous system (PNS) to sensory organs and to responsive structures, called effectors, such as muscles and glands.

The brain can be broadly divided into three regions, as in other vertebrates: the forebrain, hindbrain, and midbrain. Vision and touch are the most important senses for a sea horse. Being an ambush predator, the sea horse remains more or less stationary until prey come within range. The eyes are large, and they move separately, so the sea horse can scan a wide field of view without moving its body and betraying its presence. It swivels its eyes forward to point its snout toward individual items of prey. Then, carefully judging distance with its binocular vision, it sucks prey into its snout.

In sea horses, there is a depression, or fovea, in the retina (light-detecting layer) lining the back of the eye. The fovea contains an unusually high concentration of color-detecting cells, or cones. These cells can detect fine detail and subtle color tones—useful for an animal that selectively feeds on small prey. The sea horse's color vision also enables it to sense the color tones of the background environment and alter its skin color to match. It also reacts rapidly to the color of a mate, or an adversary, responding by changing skin color to attract or deter.

IN FOCUS

Clicking sea horses

Sea horses produce loud clicks by rubbing their bony plates together, particularly the plates in the head. Scientists speculate that the clicks may be a means of communication. Sea horses can hear one another even if they cannot see one another amid the seaweed or sea grass in which they live.

▲ *Sea horses' eyes move independently of each other, but these fish do have good binocular vision when both eyes look forward.*

Like most other fish, the sea horse probably has reasonable hearing and can detect vibrations in the water using a lateral line system. The sense of touch comes into its own when visibility is poor. Touch is also important in greeting and courtship rituals, when pairs of sea horses swim alongside each other and entwine their tails.

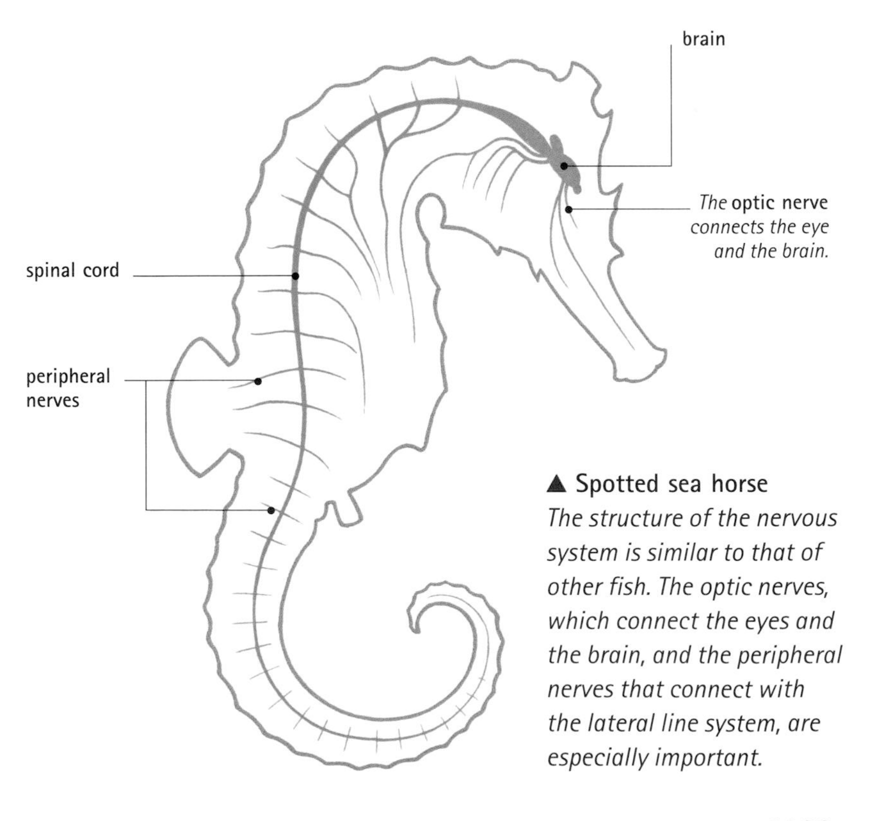

▲ Spotted sea horse
The structure of the nervous system is similar to that of other fish. The optic nerves, which connect the eyes and the brain, and the peripheral nerves that connect with the lateral line system, are especially important.

Circulatory and respiratory systems

CONNECTIONS

COMPARE the gill structure of a sea horse with that of a ***TROUT***. The sea horse's gills are composed of bulbous tubes that resemble bunches of grapes, whereas the gills of the fast-swimming trout are more efficient and are shaped like leaves.

Like other bony fish, sea horses have a single-circuit blood circulation, with blood pumped at high pressure from the heart through the gills and then on to body tissues. The blood then returns to the heart at very low pressure. Major blood vessels called arteries carry blood away from the heart, and major veins return it. Arteries divide into much smaller smaller blood vessels, called capillaries, in the gills and in all the other body organs and tissues, to allow gases and dissolved substances to diffuse between the blood and the respiring body cells of the tissues.

Four-chamber heart

The heart of a sea horse, like that of most ray-finned fish, has four chambers in sequence: the sinus venosus, atrium, ventricle, and bulbus arteriosus. Anatomically, only the atrium and ventricle are strictly true heart tissues; the sinus venosus and bulbus arteriosus are derived from sections of major veins and the main artery, the aorta. The sinus venosus bulges and releases blood smoothly into the atrium at the beginning of the heartbeat. The atrium then contracts to raise blood pressure to supply the ventricle, which is much more muscular than

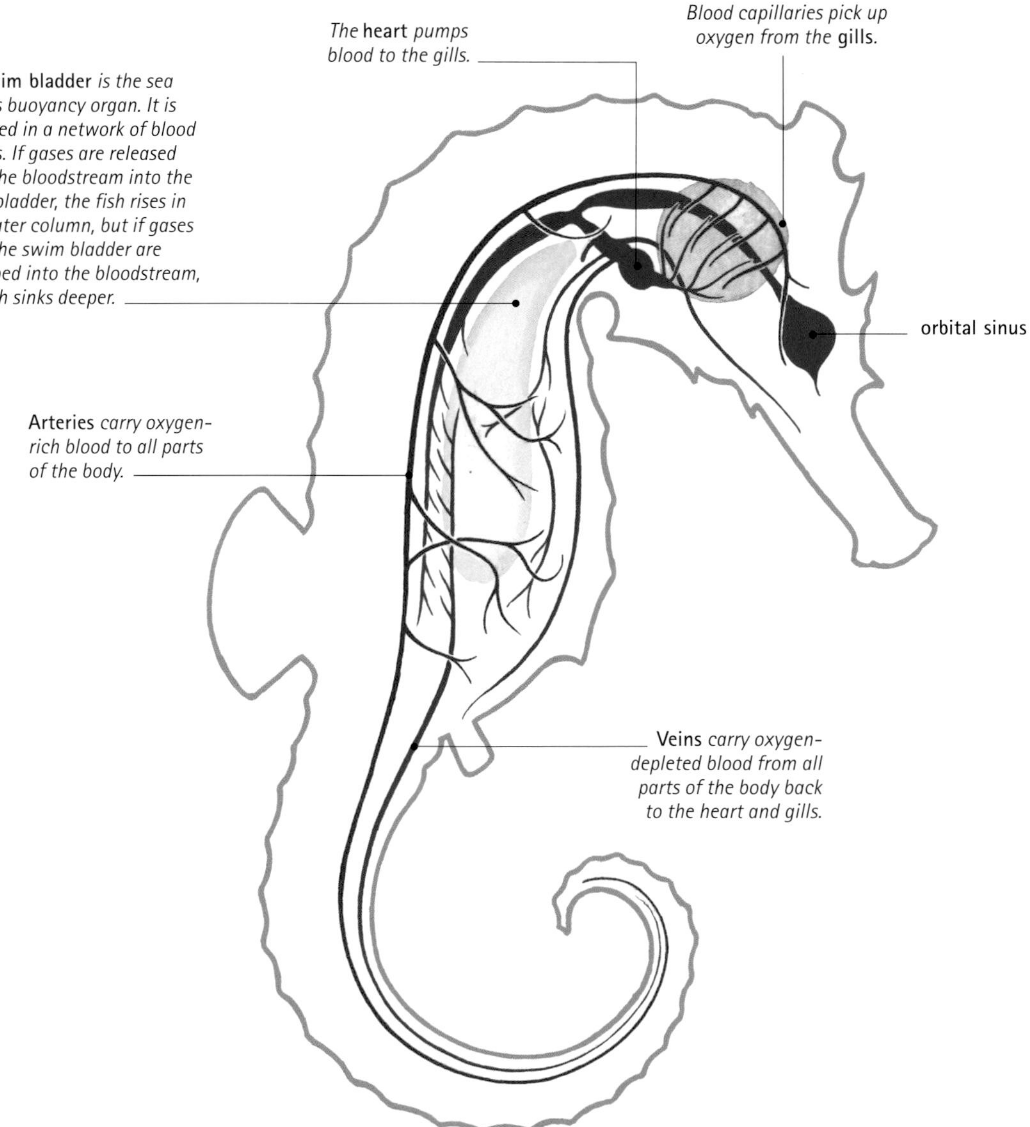

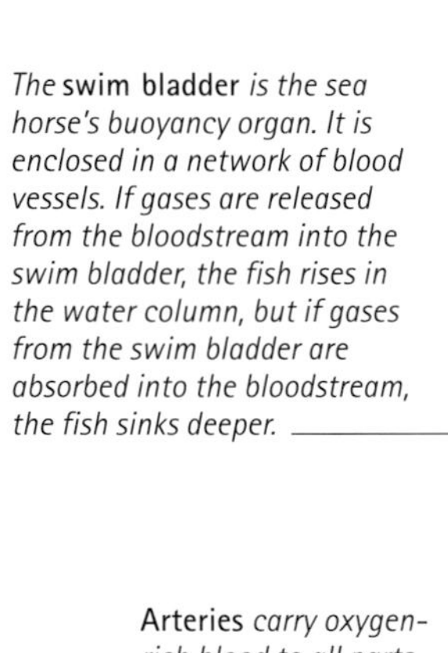

▶ **Spotted sea horse**
The sea horse has a single-circuit circulation. Blood is pumped by the four-chamber heart through the gills—where the blood picks up oxygen—and then around the body, before returning to the heart. The arterial system is depicted in red, and the venous system is shown in purple.

the atrium and raises blood pressure much higher. This is the heart's main power stroke. The bulbus arteriosus is an elastic region of the heart that dampens down changes in blood pressure from the ventricle so that the blood is pumped smoothly around the body without jerky stops and starts.

Gas exchange

Like most other types of fish, sea horses use gas exchange across the gills to gain the oxygen they need for respiration and to get rid of waste carbon dioxide. Because sea horses are slow-swimming fish, they do not have a high demand for oxygen and therefore do not need to have gills with a very large surface area, unlike fast-swimming fish, such as sailfish and trout. Sea horses' gills are unusual in resembling bunches of grapes, and the water that moves through the tubelike channels surrounded by the blood-rich tissue of the gills provides sufficient oxygen to meet the sea horse's needs.

The swim bladder

Like the swim bladders of other fish, that of sea horses has evolved over millions of years from an outfolding of the gut that has since closed off and become a separate sac. The swim bladder is enclosed in a network of blood vessels. These release gases from the bloodstream into the swim bladder or absorb gases from it, thus regulating the amount of gas in the swim bladder. To allow the fish to rise in the water, more gas is released into the swim bladder, giving the fish more buoyancy. To enable the fish to sink, gas is absorbed from the swim bladder, making the fish less buoyant. When the fish floats effortlessly at the same level in the water, it is neutrally buoyant.

In comparison with most fish, pipefish and sea horses have taken the design of the swim bladder one step further. They can adjust the amount of gas in the front part of the swim bladder relative to the rear. Doing so allows them to alter their trim, so they can float upright or tilted forward with the minimum of effort.

CLOSE-UP

Sea horse gills

Unlike other fish, sea horses have gills that are small and compact with a grapelike appearance. They look like bunches of grapes on stems rather than the leaflike arrangement of gills as found in most fish. The gills, rather than being stacked, flat plates of tissue across which water flows, are composed of bulbous tubes through which water flows. The design of the sea horse's gills is probably related to the narrow shape of the head, which is encased in bony plating, with a relatively small exit for each gill chamber.

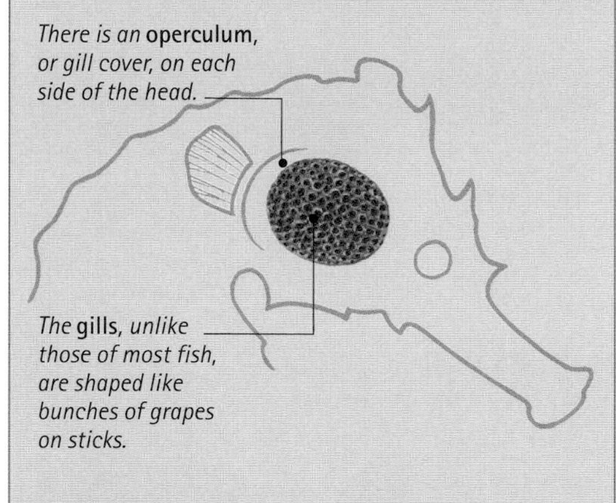

▲ GILLS
Spotted sea horse
The gills lie under the bony plate, with a small gill opening to let in water. The gills have an unusual structure that resembles bunches of grapes.

▶ *Sea horses, such as these courting tiger tail sea horses, swim slowly and therefore do not need much oxygen. This small demand for oxygen is reflected in the unusual structure of their gills.*

Digestive and excretory systems

CONNECTIONS

COMPARE the snout of a sea horse with the jaws of a ***GULPER EEL***. The sea horse's snout is a tube formed from fused toothless jaws and is used to suck in prey, whereas the gulper eel has enormous, hinged toothed jaws and a cavernous mouth for catching prey.

When it is feeding, a sea horse sucks in its prey at high speed. One sea horse kept in captivity was observed consuming 3,000 brine shrimp in a day. When a small crustacean swims within range, the sea horse points its snout toward the prey. Then the sea horse closes its gill covers (opercula), raises the roof of its mouth cavity (palate), and lowers the floor of the mouth cavity. The overall effect is to increase the space inside the mouth, which causes a drop in the water pressure inside the mouth. Water rushes in through the mouth to equalize the pressure. The prey is sucked in, along with the surrounding water. The whole process takes only 5 to 7 milliseconds (5 to 7 thousandths of a second), making the sea horse one of the fastest-feeding fish.

Sea horses have no teeth. They also have little or no stomach, as is characteristic of other fish that feed little and often on small items of prey. Swallowed food travels along the esophagus straight from the mouth to the first part of the intestine. The intestine gradually digests swallowed food, breaking down the various components of food—carbohydrates (sugars), proteins, and fats—into small molecules that can be absorbed into the blood. The digested food is delivered to body tissues, where it is used to make body parts or it is broken down to release energy in the process of respiration. Any undigested material still remaining in the intestines is voided from the body through the anus.

Hepatopancreas and excretion

In common with most other bony fish, the sea horse has a hepatopancreas, a combined organ that contains both liver and pancreatic tissue. The liver produces bile that squirts into the first part of the intestines, where it helps break down fats. The liver also stores blood sugar in the form of glycogen, which serves as an energy

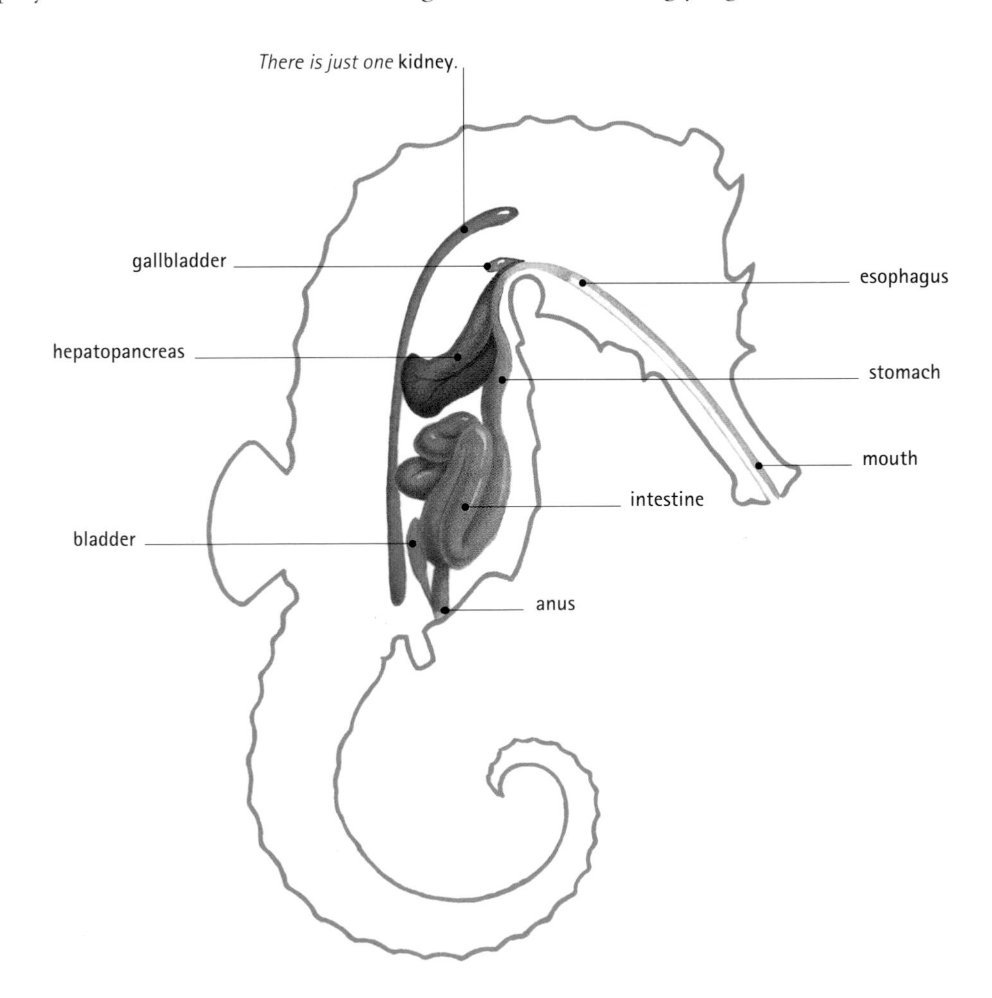

▶ Spotted sea horse
Swallowed food passes down the esophagus into the intestine, where it is digested and absorbed. Fluids produced by the hepatopancreas (combined liver and pancreas) aid digestion. Sea horses have a single kidney that excretes waste products of metabolism.

store, and it breaks down toxic substances that are circulating in the blood. The pancreas releases pancreatic juice into the first part of the intestines, where it digests fats, carbohydrates, and proteins in the consumed food. The pancreas also releases hormones (chemical messengers) that control the uptake of blood sugars, such as the glucose, which is taken up and converted to glycogen by the liver.

Single kidney

The sea horse's excretory system contains a single kidney rather than the two found in most other bony fish. Just as in other fish, the kidney excretes waste products from the blood and produces urine. In sea horses, however, the kidney is unusually simple in structure. This simplicity is all the more surprising in the few sea horses that live in freshwater. These fish have to produce large volumes of urine to expel water that passes into the body by diffusion. Scientists do not understand yet how freshwater sea horses achieve this regulation of water. In all sea horses, the urine is stored in the urinary bladder before it is expelled from the body into the environment.

PREDATOR AND PREY

Hunters and hunted

Sea horses are small but voracious predators. They consume only live prey and are opportunistic, taking most types of small floating animals that drift within range of their sucking snout, including fish fry and small crustaceans. Scientists believe that sea horses have relatively few predators, and there are several reasons for this: sea horses have effective camouflage, a secretive lifestyle, and a bony exoskeleton armed with spines and other projections. However, sea horses have been found in the stomachs of a wide variety of animals, ranging from crabs and turtles to large fish and seabirds. Young sea horses, which are not yet skilled at concealing themselves, are particularly vulnerable. Humans remain a major predator of sea horse populations, particularly in parts of Asia where the fish are dried and sold as curios or are crushed to make traditional medicines.

IN FOCUS

Different shapes of snout

The shape of the snout is the characteristic that best distinguishes different species of sea horses. The length of the snout and the size of the mouth opening reflect the size and type of food the sea horse prefers. Those sea horses that probe small gaps in coral to find food have a longer snout. Sea horses that feed on larger prey have a wider mouth opening. On a single coral reef, several species of sea horses can coexist, each species eating different kinds of crustaceans and living in distinct places on the reef.

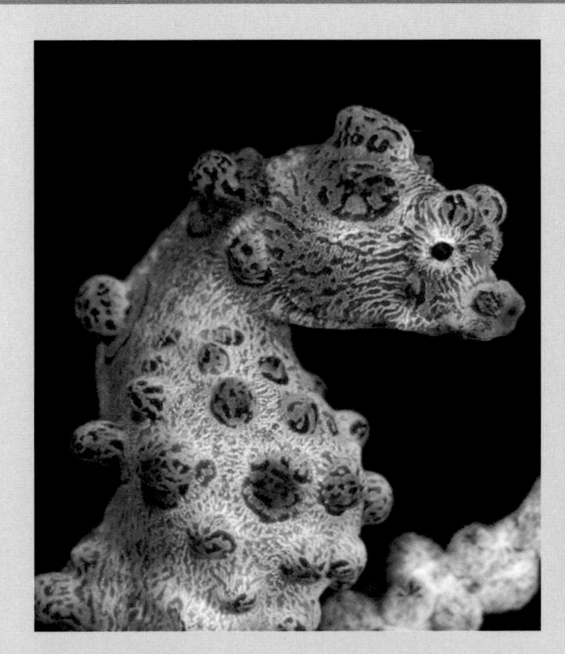

▲ *The pygmy sea horse has an extremely short snout. This sea horse preys on plankton and has excellent camouflage, mimicking the color and shape of the gorgonian coral in which it lives.*

▲ *The long-snout sea horse is able to probe nooks and crannies of the coral reef for small, swimming crustaceans and fry (immature fish) to suck into its narrow snout.*

Reproductive system

> **CONNECTIONS**
>
> COMPARE the sea horse's brood pouch with a *KANGAROO*'s marsupium and a *RAT*'s placenta. The sea horse's pouch, like a rat's placenta, provides protection, oxygen, and nutrients and removes wastes. A female kangaroo's pouch, or marsupium, protects the joey, which is nourished with milk from a teat in the pouch.

Sea horses are unusual among fish in that a male and female pair for at least one season, and often for life. Sea horses and some species of pipefish are unique among fish in that males have a brood pouch in which the female's eggs develop and hatch.

Sea horses in tropical waters can breed at any time of the year, and males without a partner fight each other for a mate. They butt each other with their long snouts, like two jousting knights on horseback. Eventually one backs down, and the victor will then court the female. Sea horses are usually territorial; they live in a restricted area and wait for drifting food to come to them. Establishing a strong mating partnership, and greeting each other at the beginning of the day, ensures that individuals do not need to move away from their territory to find their mates. In addition, the daily meeting ritual helps ensure that the pair will synchronize their breeding condition so that by the time the female has well-developed eggs, the male's pouch is ready to receive them.

The male's sperm-producing organs, called the testes, are a pair of baglike structures in the abdomen. Unusually, they produce relatively few sperm at a time: millions, rather than the billions that most fish produce. The egg-producing organs of the female, called the ovaries, are tubes that release surprisingly few eggs—only up to a few hundred at a time. This again is far fewer than in most other fish. Trout, for example, produce thousands, and Atlantic cod produce millions. Because eggs are released by the female into the male's brood pouch, where they are fertilized, far fewer sperm and eggs are needed than in fish that fertilize eggs outside the body. Eggs are protected and nurtured in the male's brood pouch, so a much greater proportion of eggs will survive to hatch.

▼ TRANSFERRING EGGS

The female spotted sea horse transfers between 20 and several hundred eggs into her partner's pouch, using a tubelike appendage called an ovipositor.

▼ GIVING BIRTH

When the young sea horses have hatched and are ready to emerge, the male sea horse contorts his body and gives birth. It may take several hours to release all the young from his pouch.

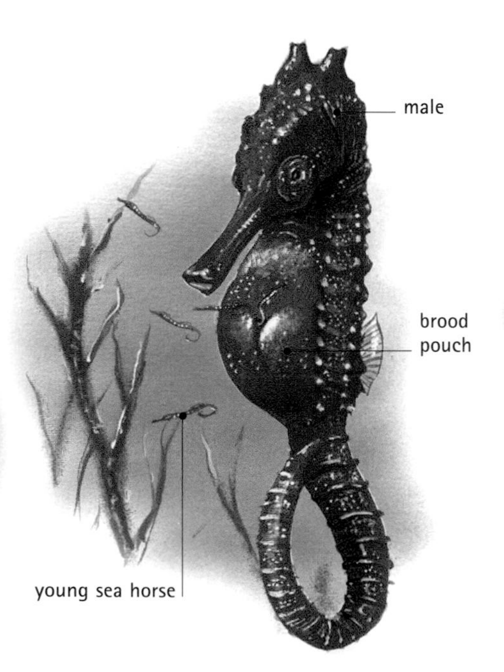

▼ INDEPENDENCE

Young sea horses cling to one another with their tails after birth, probably to avoid becoming dispersed by currents. Later, they attach themselves to corals or weeds.

Elaborate courtship

When sea horse partners greet, they perform a dance, intertwining their tails and swimming side by side in perfect unison like two ballroom dancers. The dance may be accompanied by synchronized changes in body color. When sea horses are courting just prior to egg transfer, the dance ritual becomes even more elaborate. Courtship may be spread out over several days, until the female is ready to lay her eggs in the male's brood pouch. To do so, she entwines her body around his and inserts her ovipositor, an egg-laying tube, into the single opening near the base of the male's pouch. She squirts the eggs into his pouch, where he fertilizes them.

In many species of sea horses, including the spotted sea horse, the brood pouch seals up to protect the eggs. Tissue grows to surround the eggs and they become bathed in a nutritive fluid that is kept supplied with oxygen by a network of small blood vessels in the wall of the pouch. The brood pouch provides protection, oxygen, and nutrients, and removes wastes, thus doing the same job as the placenta of a pregnant female mammal. Because the developing fish are nourished by the father, they do not have to rely just on egg yolk for the energy and raw materials to grow. At birth, sea horses are well developed and can lead an independent life.

IN FOCUS

Egg protection

Between them, pipefish and sea horses show a range of strategies for caring for eggs and young. At one extreme, the eggs of some pipefish are simply loosely attached to the dorsal surface of the male. In others, the eggs are embedded in spongy tissue on the skin surface. In some pipefish and all sea horses, the eggs are partially or completely enclosed in a chamber. Pipefish in the genus *Syngnathus* have a brood pouch located on their back, or dorsal surface, whereas sea horses have a brood pouch on the ventral surface (underside).

▼ Male spotted sea horse

The male sea horse's baglike testes produce sperm. The female deposits eggs into the male's brood pouch, where they are fertilized and nurtured.

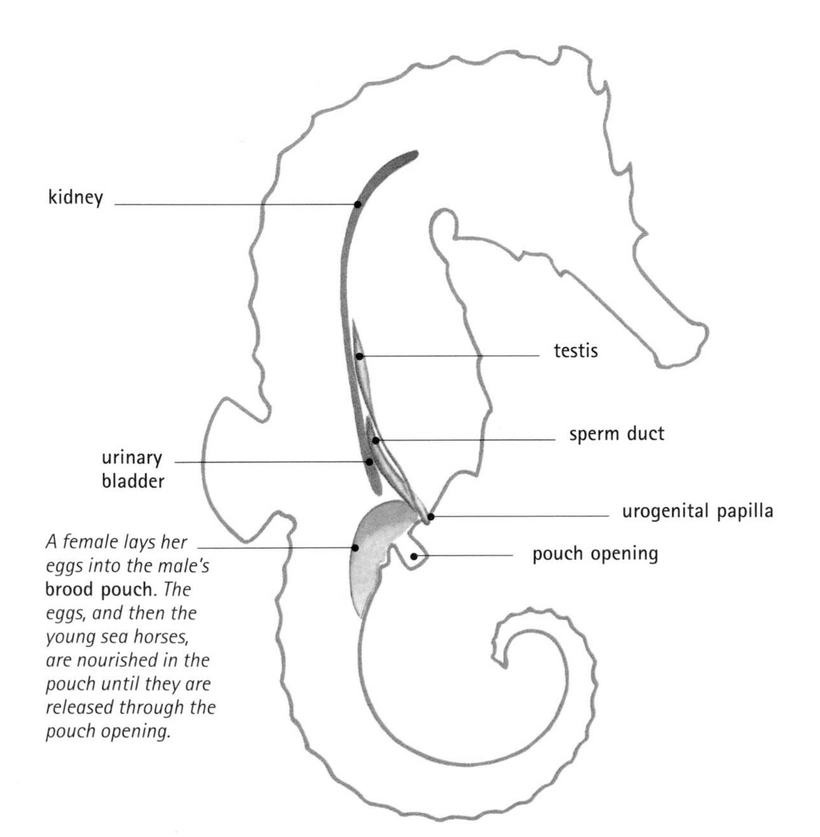

A female lays her eggs into the male's brood pouch. *The eggs, and then the young sea horses, are nourished in the pouch until they are released through the pouch opening.*

Development and birth

Depending on the sea horse species and the water temperature, the eggs hatch inside the pouch after two to six weeks. In most sea horse species, the balance of water and salt inside the pouch is gradually altered until it becomes more and more like that of the surrounding seawater into which the young will be born. This change minimizes the physiological shock as young sea horses leave the pouch.

At birth, the male contracts the abdominal muscles to squeeze on the pouch and pump the young sea horses out through the pouch opening. Sea horses are born at 0.25 to 0.5 inch long (0.6–1.2 cm) and resemble miniature versions of the adults. Birth usually occurs under cover of darkness, so the newborn fish stand a good chance of avoiding predators until they find a place to hide. Afterward, the male flushes his brood pouch with water to expel any egg fragments and other remains. In most species, neither parent plays any role in caring for the young after they are born. In most sea horse species, the young grow to maturity by the end of the first year and live for between three and five years in the wild. In captivity, they may live longer.

TREVOR DAY

FURTHER READING AND RESEARCH

Kuiter, R. H. 2000. *Seahorses, Pipefishes, and Their Relatives: A Comprehensive Guide to Syngnathiformes.* TMC: Chorleywood, U.K.

Petrinos, C. 2001. *Realm of the Pygmy Seahorse: An Underwater Photography Adventure.* Starfish: Athens, Greece.

Seahorse.org: www.seahorse.org

Seal

ORDER: Carnivora FAMILIES: Phocidae, Otariidae, and Odobenidae SPECIES: 33

Centuries ago, sailors called seals "sea bears," and bears are indeed seals' closest living relatives on land. Both bears and seals are members of a group of mammals called the carnivores, or meat eaters. The anatomy of seals reflects the fact that they spend most of their life in water, so their eyesight, hearing, body shape, and method of movement all work better in water. However, seals' anatomy is also affected by their ancestry as air-breathing land mammals, and they do not live all their life in water. Seals must leave water and stay on solid surfaces to rest, breed, and molt their skin. Therefore, they must be able to move about, see, and hear adequately in air as well as water.

• **Animals** Animals are many-celled organisms that actively eat other organisms to obtain energy and nutrients. Most are responsive to external stimuli and have cells that are organized into tissues and organs.

▼ *There are three families of seals: true, or earless, seals form the largest family, which includes the harbor seal and the ringed seal; there are 14 species of fur seals and sea lions; and the walrus is in a family of its own.*

• **Chordates** Chordates are animals that, in at least one stage in their life cycle, have both a dorsal nerve cord and a notochord running along their back. The nerve cord is a bundle of nerve fibers, and the notochord is a stiff rod that in most chordates develops into a backbone.

• **Vertebrates** Chordates that have a notochord that changes into a backbone during the development of the embryo are called vertebrates. They include fish, reptiles, amphibians, birds, and mammals. A backbone, or vertebral column, comprises a chain of small units called vertebrae, which are made of cartilage or bone. Vertebrates also have a braincase, or cranium, which gives the group the alternative name Craniata.

• **Mammals** Mammals are vertebrates in which females have mammary glands that secrete milk to feed the growing young. Most mammals have fur. They all have a jaw that hinges farther forward than that of their reptilian ancestors, and the lower jaw comprises a single bone, the dentary. The teeth have diverse forms and functions within the mouth, and the teeth mesh together precisely and can grind food. Mammals' mature red blood cells lack a nucleus, unlike those of reptiles and birds.

• **Carnivores** Mammals that are usually meat eaters, such as cats, dogs, badgers, weasels, bears, raccoons, and seals, make up the order Carnivora. Most members of the order have bladelike cheek teeth called carnassials, which slice the food before swallowing. Not all meat-eating mammals

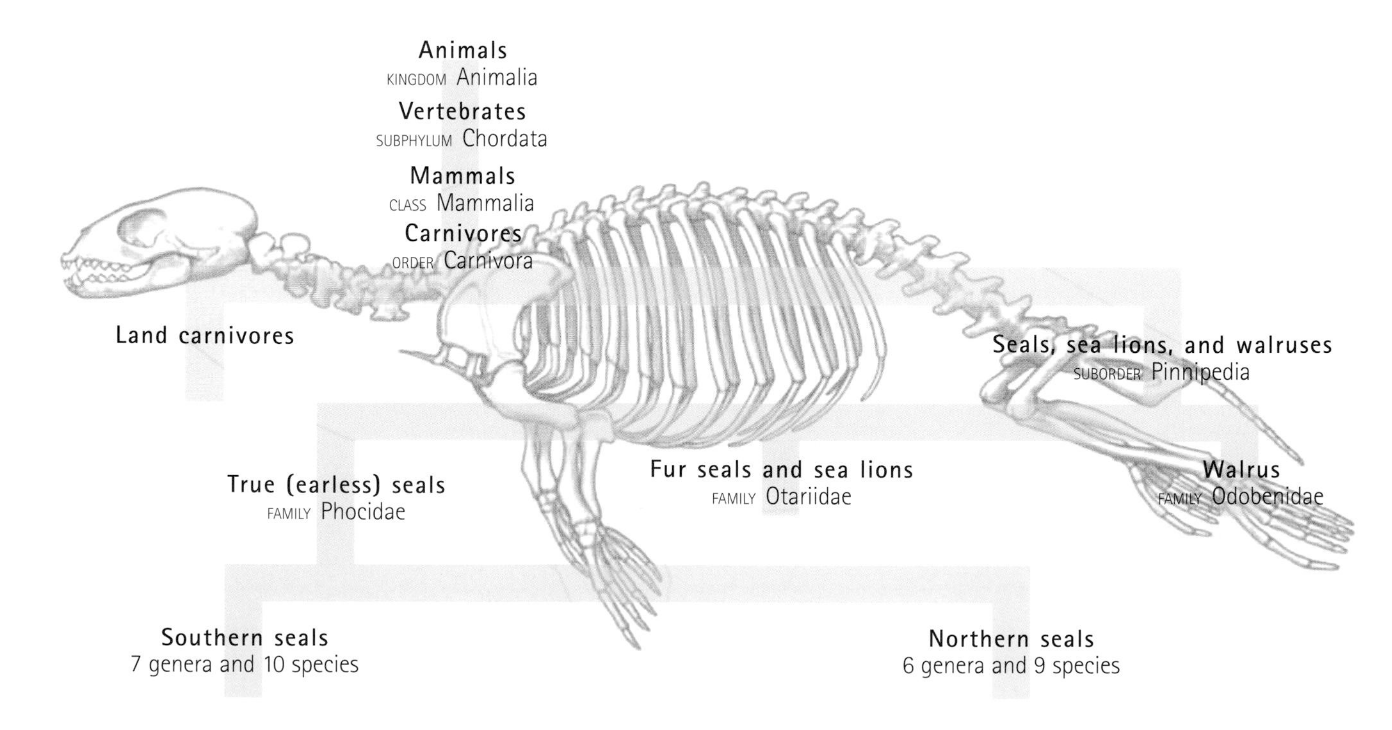

are in this order, and not all members of this order eat meat. Bears and raccoons eat both plants and animals, and the panda eats only plants. Seals do not have the carnassial teeth characteristic of most animals in the order Carnivora.

▲ *This Weddell seal is searching for prey. Like all seals, it has a body shape perfectly suited to rapid movement underwater.*

• **Pinnipeds** Seals, sea lions, fur seals, and walruses are marine carnivores with paddle-shape flippers that are used for propulsion in water. The pinnipeds, whose name means "finned foot," include the eared seals (sea lions and fur seals); earless, or "true," seals (such as the ringed seal); and walruses. Pinnipeds have a body that is covered with a thick layer of fat, giving the body a neat, streamlined shape and insulating the internal organs from cold water.

• **Eared seals** Sea lions and fur seals are pinnipeds with small external ears. They look different from the true seals because they are more mobile on land and are able to hold their body more erect, sometimes raising it clear of the ground using both foreflippers and hind flippers.

• **Walrus** There is only one living species of walrus. This huge pinniped lives in the Arctic and has very sparse fur and enormous tusks.

• **True seals** These pinnipeds have no external ears. True seals are more fully suited to life in water than the eared seals. On land, true seals cannot bring their hind flippers forward beneath their body to raise it above the ground. Their forelimbs are buried in their thick fat layer up to the wrist, so the limbs appear very short. True seals move on land by an undulating movement, but in water they are much more accomplished divers than eared seals.

• **Northern true seals** The northern true seals differ from the southern branch of the family (the monk, elephant, and Antarctic seals). The northern seals' flippers have long, thick claws, but southern seals' claws are small and do not extend beyond the end of the flipper. Southern seals' hind flippers are thickened and stiffened with fibrous tissue and look a little like the tail flukes of dolphins. Northern seals' hind flippers are more flexible and can bend to scratch each other.

• **White-coated true seals** The harbor seal belongs to a small grouping of northern true seals whose young are born with a dense, white fur coat called lanugo. This group also includes the ringed seal, harp seal, and gray seal.

FEATURED SYSTEMS

EXTERNAL ANATOMY Although some seals look clumsy on land, their body is streamlined and cuts easily through water, propelled by flipper-shape limbs. *See pages 1112–1115.*

SKELETAL SYSTEM The skeleton of seals does not offer much support to the body. The limb bones are short, sturdy appendages with long digits forming paddles. The flexible spine allows underwater maneuverability. *See page 1116.*

MUSCULAR SYSTEM A seal's muscles act as an important oxygen store. They are packed with myoglobin, a protein that binds spare oxygen and releases it as required when the seal is diving. *See page 1117.*

NERVOUS SYSTEM Seals must hunt in dark polar winters and find food deep in the ocean beyond the reach of the sun. Their vision is excellent, and hearing, touch, and taste are all highly developed, too. *See pages 1118–1119.*

CIRCULATORY AND RESPIRATORY SYSTEMS Seals are champion divers, equaled in duration and depth only by the largest species of whales. Their circulatory and respiratory systems are suitable for operating under high pressure and for long periods of time without inhalation of air. *See pages 1120–1121.*

DIGESTIVE AND EXCRETORY SYSTEMS A seal's diet consists exclusively of fish and shrimp, and thus lacks both carbohydrate and freshwater. The seal is able to process the fat and protein in its food to release all the water and carbohydrate it needs. *See pages 1122–1123.*

REPRODUCTIVE SYSTEM By producing rich, fatty milk for their young, female seals transfer their protective blubber to their offspring; as a result, seal pups have a very fast growth rate. *See pages 1124–1125.*

External anatomy

CONNECTIONS

COMPARE the sleek body shape of a seal with that of another warm-blooded diver in cold seas such as a ***PENGUIN***.

COMPARE the touch-sensitive hands of a ***MANDRILL*** with the touch-sensitive whiskers of a seal.

Most seals are mammals of cold oceans. Seals live farther north and south than any other group of mammals. In the far north, ringed seals live under the permanently frozen surface of the Arctic Ocean and carve snow lairs where they find weak points in the ice. Ringed seals swim between breathing holes 0.6 mile (1 km) apart with ease. In the far south, the Weddell seal gnaws the sea ice to maintain breathing holes that allow it to live far under the permanent ice around Antarctica. It can swim 3 miles (5 km) away from its breathing hole before returning. The water in its habitat has a temperature around 29°F (–1.8°C), but on the ice shelf the air temperature can drop to –58°F (–50°C) and howling gales produce an intense wind-chill factor. Since they are mammals, all seals nevertheless maintain a core body temperature of 99°F (37°C).

Not all species of seals live in extremely cold environments. Most seals prefer water below 68°F (20°C), but some, such as the Hawaiian monk seal, live in tropical seas. Even so, the body form of seals is dominated by the need to swim well and stay warm. Therefore, all seals have a smooth, streamlined shape that slips through the water with minimal turbulence and drag. Seals' large size and simple shape minimize body surface area relative to mass

▶ Harbor seal

The harbor seal has a broad, rounded, doglike head, no visible ears, large eyes, a streamlined body, and short flippers. The fur of adults and pups is pale to dark gray.

▲ *Two Weddell seals have swum up to a hole in the ice to breathe air. Individuals of this species sometimes spend more than 80 minutes underwater, but they have to breathe air periodically. The seal in the foreground is a pup.*

and volume; this arrangement serves to minimize heat loss. The ringed seal is the lightest pinniped, at about 110 to 210 pounds (50–95 kg). Many pinnipeds are much larger. The elephant seal, for example, grows to a massive 8,000 pounds (3,600 kg).

Seals have curved contours because they are covered by a thick layer of fat called blubber, which smooths all their sharp corners. Any features that would spoil a smooth outline, such as nipples and genitals, are tucked away in neat grooves. Fur seals and sea lions (together called eared seals) have small external ears, which cause a little water resistance. The more aquatic true, or "earless," seals, such as the harbor seal, have no external ears at all.

Blanket of fat

The blubber layer of seals serves several purposes; one of the most important is the conservation of heat. Blubber forms a blanket

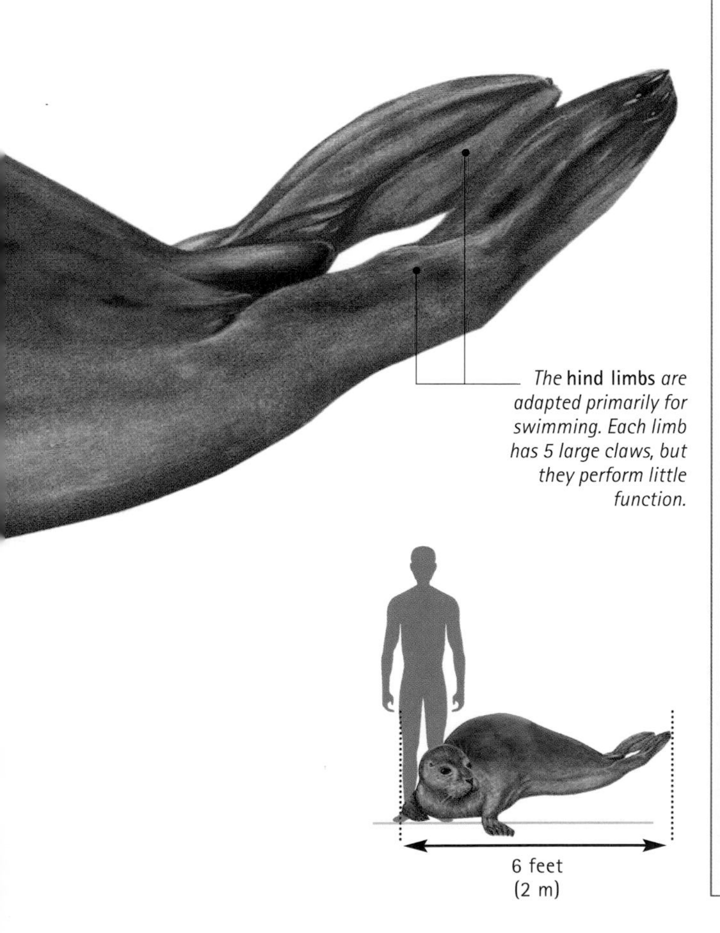

The **hind limbs** *are adapted primarily for swimming. Each limb has 5 large claws, but they perform little function.*

EVOLUTION

Pinniped origins

Scientists still argue about pinnipeds, but most now agree that all pinnipeds descend from the same land-based, bearlike ancestor, which took to the seas. A fossil animal from California sheds some light on pinniped evolution. Named *Enaliarctos mealsi*, it lived 23 million years ago. Some scientists believe it represents a snapshot of the evolutionary change from bearlike carnivore to a seal-like animal. *Enaliarctos* swam with undulations of its body, like a seal, but it also propelled itself with thrusts from its fore flippers, like a sea lion, and strokes from its hind limbs. The combination was not unlike the swimming of today's otters, and the artist's impression below reconstructs the animal as otterlike in appearance.

▼ *Enaliarctos mealsi*
The ancestor of seals was related to bears but probably had an appearance more like that of an otter.

▲ *This harp seal pup is molting its fur. White juvenile fur remains on the posterior (rear) section of the body, and gray adult fur has grown on the forequarters.*

of insulating material up to 4 inches (10 cm) thick in walruses and elephant seals. This insulation is crucial for seals, because water has a high heat capacity: that is, it saps heat energy from the seal much more quickly than does air. A temperature gradient exists from the seal's warm body just inside the blubber to the cold skin outside. Seals keep their skin at just 34°F (1°C) when lying on ice, just warm enough to avoid freezing. Their skin cells are much more tolerant of cold air and water than are those of other mammals.

Sealskin is not shed flake by tiny flake as it is in humans, because the dead skin would become waterlogged. Instead, seals molt their skin in one go. Elephant seals' skin, in particular, comes off in great patches several inches across. During the molt, seals remain out of water and thus are unable to feed.

CLOSE-UP

Big eyes for dark seas

The eyes of seals have pupils that can become as narrow as pinpricks for viewing polar ice in bright sunshine. Such tiny pupils let in little light. The pupils can also widen greatly to let in the maximum light during hunting deep in the sea, at night, or beneath the sea ice. Seals that live near the poles endure long months of darkness when the sun does not creep above the horizon in winter. Seals produce thick, viscous tears to protect their eyes from salt water. However, they have no tear ducts to channel the tears away, so on land, seals always have tears streaming down their face.

Buried limbs

The blanket of blubber covering a seal's body buries the base of its appendages. The limbs of true seals emerge only at the wrist or ankle. Their forelimbs are short and shaped like flippers. The long digits of the "hands" are fully webbed with fibrous connective tissue, so the fingers cannot move independently. The harbor seal, like other northern true seals, has a large, sturdy claw on each digit, with which it

CLOSE-UP

Mustachioed hunter

One part of the seal's skin that is never allowed to cool is the sensitive pad above its mouth bearing a mustache of vibrissae (whiskers). This patch of skin shows up clearly on thermographs (images made with heat-sensitive cameras). Even in the coldest water, seals cannot tolerate chilling of this area, because it must remain warm to stay sensitive. Harbor seals use their whiskers to distinguish the size, shape, and surface structure of objects by touch. They can find their flatfish prey on the seabed in this way. Seal whiskers can also detect the water disturbances caused by swimming fish. Such water vibrations persist for some minutes after a fish has swum by. A hunting seal can follow the wake of the fish to find its quarry.

▼ HIND FLIPPERS

The hind flippers of seals vary greatly. For example, the claws of harbor seals are much larger than those of elephant seals.

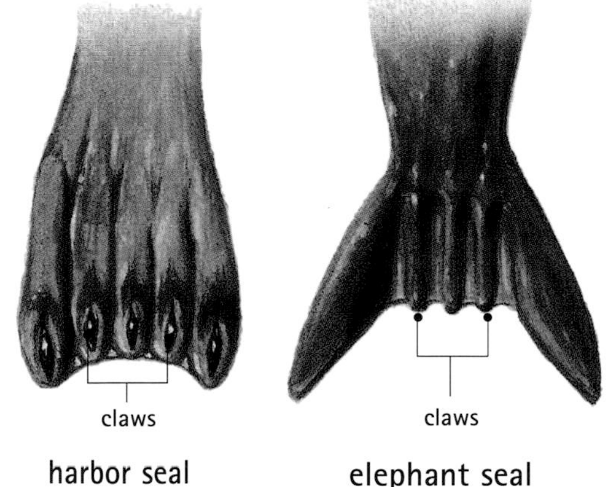

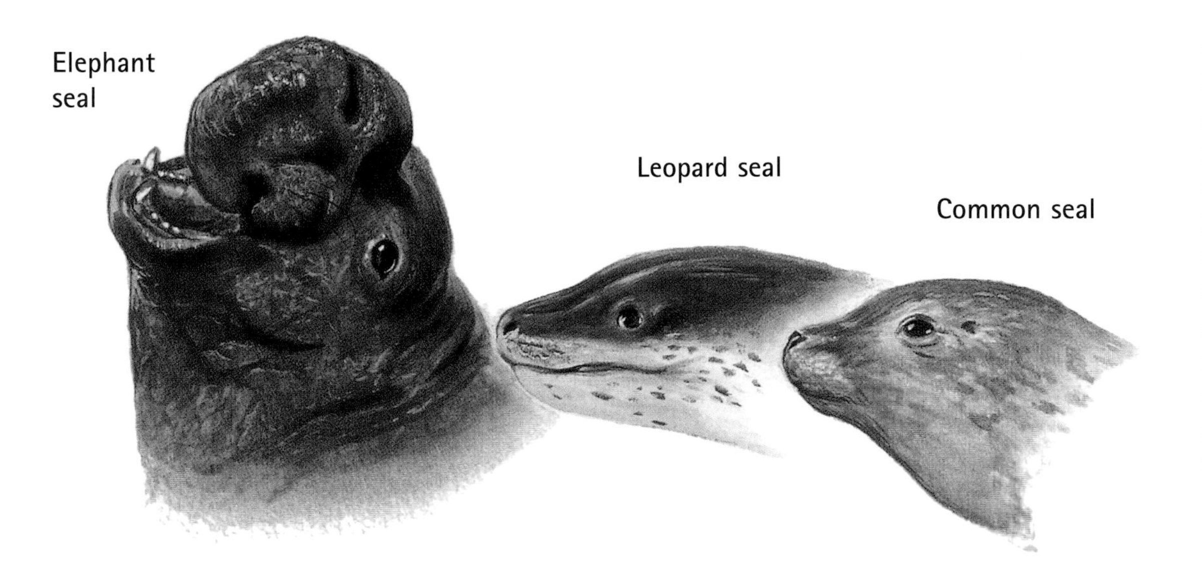

◀ HEAD SHAPES
The enlarged muzzle of a male elephant seal amplifies its roars, and the powerful, thickset head of a leopard seal is suited to preying on other seals and penguins. The rounded, doglike head and big eyes of the common seal are typical of the family.

can scratch and groom itself. The feet are similarly webbed, forming paddles. The first and fifth digits are longest, so the hind flipper has a symmetrical fan shape. The role of the hind flippers is mainly to generate propulsion through water, and they are almost useless on land. During swimming, the hind flippers act as hydrofoils, generating both lift and propulsive force, similar to the flapping wings of birds. The front flippers are held flat against the side of the body in small recesses. The seal can use its foreflippers for maneuvering or for extra propulsion in emergencies.

The flippers of eared seals are very different. The limbs that provide propulsion are the long, powerful forelimbs, and eared seals "fly" underwater like penguins, using the front flippers as hydrofoils. The forelimbs emerge from the body surface midway along the forearm, and bend backward at the wrist to support the front of the animal's body. The hind flippers can be rotated beneath the body so the toes face forward, as in land-based mammals. On land, therefore, an eared seal can support itself and walk on both its foreflippers and hind flippers. All the flippers have hairless "palms" that contact the ground.

COMPARATIVE ANATOMY

Fur versus blubber

True seals such as the ringed seal have fur, but it is nothing compared with the coats of their relatives, the fur seals. In all pinnipeds, the hair grows in units. Each unit has a long, stiff guard hair and some finer fibers underneath. Each fiber grows from its own follicle, but a unit of fibers emerges as one from the skin surface. True seals have only a few fine fibers accompanying each guard hair, or sometimes none at all. Fur seals have up to 50 fine fibers per guard hair, making a very thick coat of up to 390,000 hairs per square inch (60,000 per sq cm). Secretions of oily sebum from sebaceous glands in the skin condition the fur and make it waterproof, so it offers effective insulation in both air and water. Sodden fur would provide very little insulation. True seals dive so deep that their coat becomes compressed and much of the trapped air that provides insulation is squeezed out. Thus true seals rely on fat for insulation instead of fur and have a much thicker covering of blubber than fur seals do.

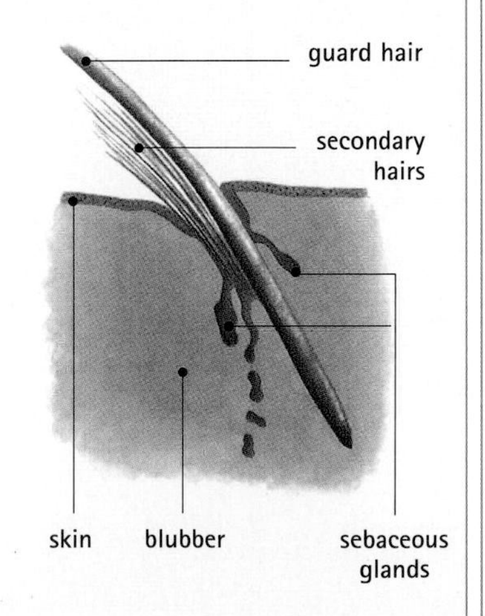

▶ FUR SEAL HAIR BUNDLE
Fur seals have up to 50 fine secondary hairs with each primary hair. This arrangement provides a very dense mat of fur. Sebaceous glands secrete an oily substance that waterproofs the fur.

Skeletal system

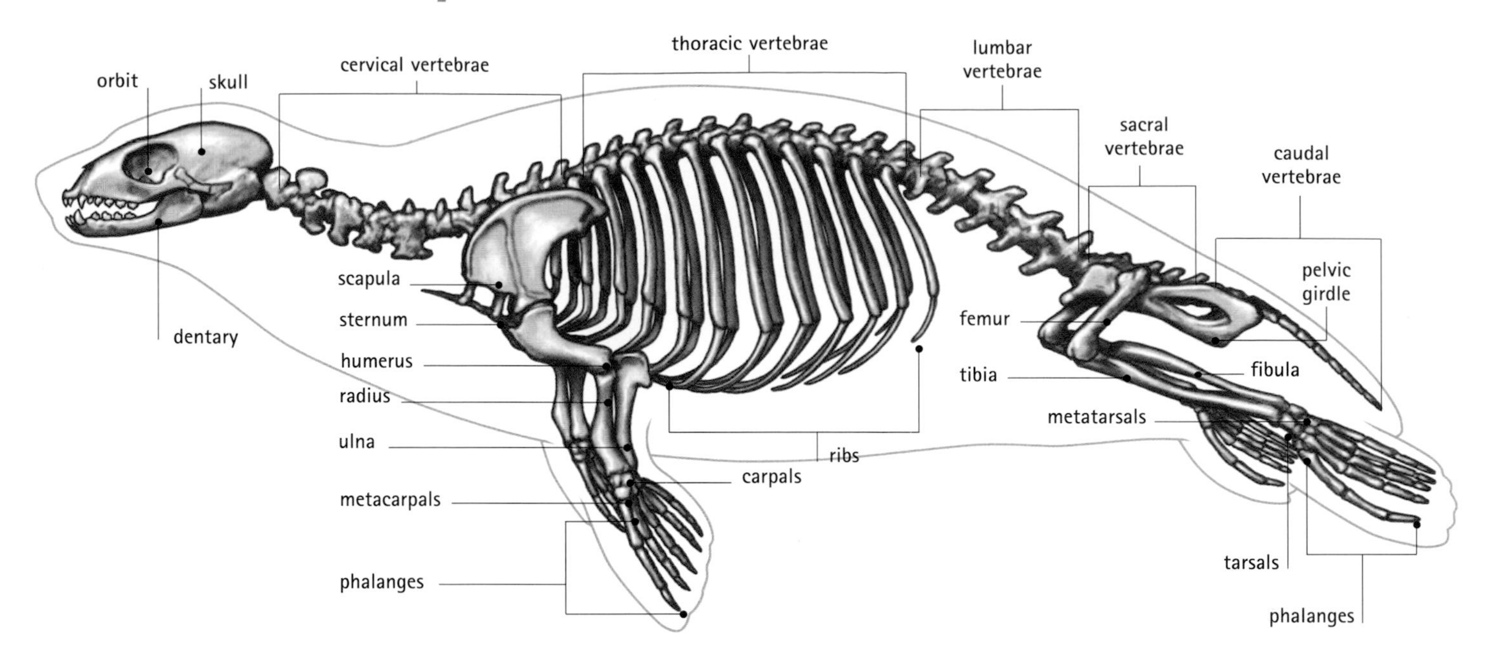

▲ **Harbor seal**
The lumbar vertebrae are large and strong. They provide anchorage for the powerful muscles that drive the seal through the water. The limb bones are short and powerful.

Under a seal's skin and soft tissue are most of the familiar bony elements of a typical mammalian skeleton. The parts that differ most from that of its land-based relatives, such as bears and dogs, are the limbs. The bones of the seal's forelimb—the humerus, radius, and ulna—are short, stout, and heavy, and are buried in the seal's blubber. Only the wrist bones and long digits are free. The limbs are not connected by a clavicle (collarbone), so they are free to move in all directions. Seals can scratch their chin, then reverse and rotate their flipper to scratch the top of their head. The phalanges (bones of the digits) are extremely long, none more so than the pollex (thumb), which forms the long leading edge of the flipper. The major bones of the hind limbs (femurs, tibiae, and fibulae) are short, broad, and flattened. A seal's femurs (thighbones) stick out sideways from the pelvic girdle (hipbone), so that although the seal's rear is narrow and tapered, its hind flippers are set wide apart.

Seal skulls are generally short, with huge orbits (the recesses that accommodate their large eyes). In the smallest species, the ringed seal, only 0.1 inch (3 mm) of bone separates the two orbits, and the bone is delicate and almost transparent. As in the skulls of other seals, the shortness of the ringed seal's snout is acheived by overlapping supraoccipital and parietal bones. A short snout increases the biting power of the jaws.

In the backbone of the ringed seal and other true seals, the lumbar vertebrae (those between the ribs and the pelvis) are the most heavily built bones. They bear sturdy spines that point sideways and serve as attachments for the swimming muscles. The spines on the tops of all the vertebrae (the neural spines) are very short, however, as are the zygapophyses (the interlocking knobs where the vertebrae link together). This gives the backbone great upward flexibility, so that, amazingly, a seal can bend backward and touch its tail with its nose.

COMPARATIVE ANATOMY

Fur seal skeleton

There are major differences in skeletal structure between true seals and eared seals (fur seals and sea lions). The most important differences are in the shoulders and forelimbs. Eared seals, unlike true seals, power themselves through water with their forelimbs, and it is the neck region of the spine that is heavily built, not the lumbar region as in true seals. All the forelimb bones are bigger and longer than those of true seals. The ridges on the huge shoulder blades and the long, broad neural spines on the neck (thoracic) vertebrae all act as attachments for powerful muscles associated with swimming.

Muscular system

The locomotion of seals, particularly true seals, is controlled by muscles very different from those that power the movement of land mammals. True seals, such as the ringed seal and harbor seal, swim by undulating the hind body from side to side. Their hind flippers face inward. They alternate in making inward-moving power strokes with the digits spread, followed by outward-moving recovery strokes with the digits curved and closed.

The main swimming power comes from long muscles in the lumbar region of the spine: the iliocostalis and the longissimus. When these muscles contract, they flex the spine to and fro. The muscles are aided by long tendons running to the flippers and by sets of muscles on each side of the limb bones: the gracilis, biceps femoris, and semitendinosus. These muscles firmly secure the hind limbs close to the pelvis and nearly parallel to the spine. With the limbs in this restricted position, the power from the lumbar muscles is effectively transmitted down to the flippers. Because of this arrangement, true seals cannot rotate their flippers forward under their body as eared seals and land mammals can.

▼ *A true seal's neck muscles are so thick and powerful that no narrowing of the neck is visible on the exterior of the seal's body.*

CLOSE-UP

Muscles for diving

Seal muscles must function when the seal is underwater and their oxygen supply is restricted. The muscle fibers (the extremely long cells of muscles) are packed with myoglobin. Myoglobin is the pigment protein that gives muscle its dark red color. It binds oxygen more tightly than the blood pigment hemoglobin, enabling the muscles to absorb oxygen from the blood and store it. Seals' swimming muscles contain between three and seven times as much myoglobin as those of land mammals. Those of Weddell seals have 10 times the myoglobin concentration of land mammals, and the muscles appear almost black when exposed to air.

Neck muscles

The neck muscles keep the seal's neck flexed in a shallow U-shape. A short, stiff neck is necessary because the seal must maintain a strong, hydrodynamic profile into the flow of water as it is powered along from behind. However, the neck muscles also permit a sudden extension of the neck to change direction or to lunge for prey.

▼ FOREGLEG MUSCLES

Sea lion

The foreleg muscles power the paddlelike forelimbs, which are all-important for propelling seals and other pinnipeds through water.

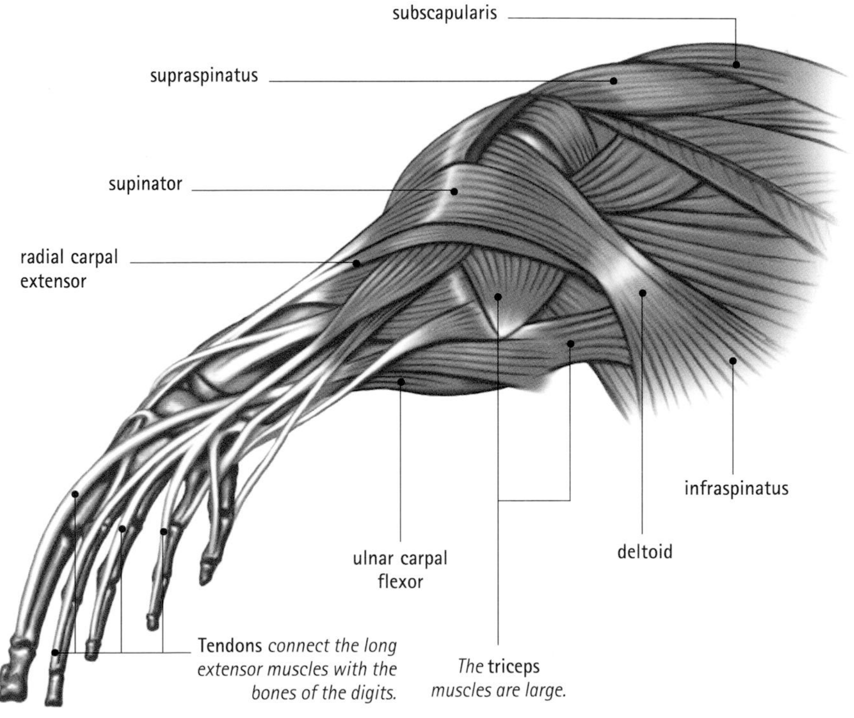

Nervous system

Like the nervous system of any vertebrate, a seal's nervous system is centered on its brain. The brain receives messages from the network of nerves branching through every part of its body. Many of the nerves receive messages from sense organs, such as the eyes and ears. The brain processes the messages and sends instructions for action to muscles.

A seal's brain is much like the brain of other carnivores, such as bears and dogs. It is more or less spherical and has a folded surface. It has a particularly large cerebellum, which is the region responsible for coordinating precise movements of the body. This gives the seal the fine control it needs to pursue fast-moving prey that move both horizontally and vertically.

Eyes, ears, and whiskers

The seal's sense organs have been principally shaped by the need to catch prey in demanding environments. The eyes, ears, whiskers, and taste buds all work well in water and in air. They guide seals in the darkness of the deep sea and in the long, dark polar winter. Seals' vision is therefore excellent and similar to that of nocturnal mammals. Seals' eyes are very large, capable of maximizing brightness and sharpness of vision in low light. Also, just behind the retina (the layer of light-sensitive cells on the back of the eyeball) is a reflective layer called a tapetum lucidum, which bounces light back toward the retina, thus making the most of whatever light is available. Many nocturnal animals have a tapetum lucidum.

Seawater scatters and absorbs light and does not allow it to penetrate to great depth. Red light is absorbed first, leaving only green and blue light, and at greater depths, blue only. Deep-diving seals, such as the elephant seal, have eyes that are most sensitive to the wavelengths of blue light, whereas seals that remain in the shallows, such as the spotted (larga) seal, are most sensitive to green light. Seals do not truly see in color, but they do have two different types of visual pigment (pigments are the light-sensitive chemicals in the cells of the retina), which are sensitive to different colors. Experts think seals use these different sensitivities to see color contrast and are thus able to see objects against a colored background.

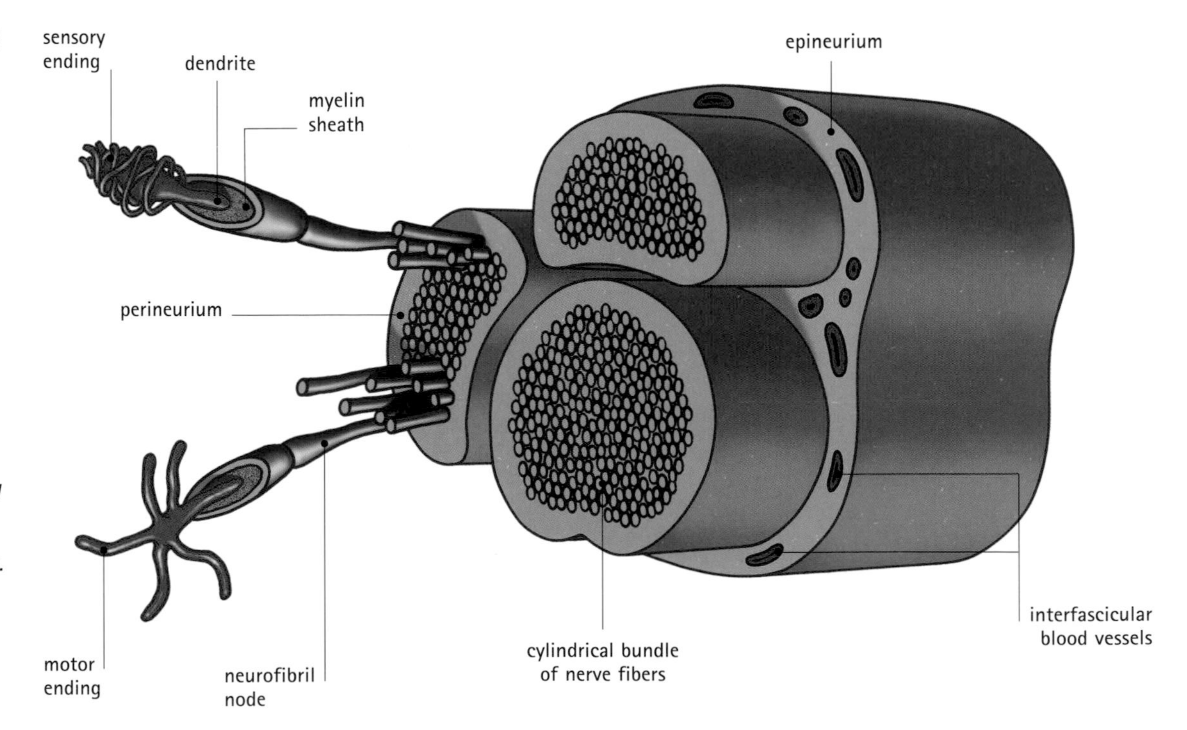

▶ PERIPHERAL NERVE *Like other mammals, seals depend on an extensive network of peripheral nerves to supply the brain with sensory information and to carry motor signals back to organs such as muscles. The nerves are insulated by a sheath of fatty tissue called myelin, which prevents their electrical signals stimulating inappropriate reactions.*

IN FOCUS

A refined palate

A seal's sense of taste was once thought to be unrefined and unimportant. However, it is now know to be finely tuned to differences in the saltiness of seawater. By tasting the water, a seal gets clues about its location and thus can home in on areas where prey are likely to be (fish often congregate in certain places relative to freshwater runoff from land). Seals can discriminate the difference between 3 percent salt in seawater and 3.1 percent or 2.9 percent salt. That sensitivity is 4.5 times better than in humans, and better than in any other mammal.

▲ *A seal has many whiskers around the mouth and nose. Each whisker sits in a hair follicle with a dense mesh of nerves. Each time a whisker moves—even slightly—nerve impulses are sent to the brain.*

Whisker stimulation

Blind seals have been known to survive perfectly well in the wild, so eyesight is apparently not vital to their survival. Other senses must compensate, and one of them might be the touch sense of the seal's whiskers, or vibrissae. Each whisker sits in a hair follicle served with a dense mesh of 1,000 to 1,600 nerves—10 times the number associated with the whiskers of cats or rats. When the whisker is deflected by an object or by water movement, it bows within the follicle and stimulates tiny sense organs that detect pressure and stretch. They send impulses to the brain along the trigeminal (facial) nerve, which is particularly large in seals. The sense of touch could help seals find their way—and even find their prey—in the dark.

Underwater hearing

Seals hear a little better underwater than they do in air, but it is not clear exactly how. In water, an external ear is transparent to sound—sound waves pass straight through because a seal's body tissues are about the same density as water. That is the reason why true seals do not need external ears. However, the sound waves also pass straight through the seal's head, striking only the bones in its skull, which are less dense. The sound is conducted through the skull bones to the middle ear, reaching and stimulating both inner ears at about the same time. It is not clear how the seal can then tell which direction the sound comes from. Nonetheless, seals have good directional underwater hearing. It may be that the route of sound into their ears is restricted. For example, in dolphins, this restriction is achieved by the isolation of the ear bones from the skull, and the channeling of incoming sound through a conductive fat channel in the jaw. A seal's ear bones—its periotic bones and tympanic bulla—are not isolated.

CLOSE-UP

Vision in air and water

In air, the eyes of mammals focus incoming light with the power of two lenses: the cornea (the curved protective front covering of the eye) and the lens itself, behind the cornea. In water, the cornea effectively disappears, because it has roughly the same refractive index as water, so it cannot bend light. The lens is still effective, though, and that of seals is large and almost spherical to compensate for the lack of corneal focusing. In air, seals should be nearsighted because the focusing power of the cornea should come into effect. However, seals have a flat cornea that may actually have negative refractive power, correcting the overpowerful spherical lens. Even so, seals' eyesight is not as good in air as it is in water.

Circulatory and respiratory systems

An animal's circulatory and respiratory systems work together to obtain oxygen and to pass it around its tissues. For an air-breathing animal such as a seal, the respiratory system comprises a pair of lungs, a diaphragm, and airways in the nose and throat. The circulatory system (the heart and blood vessels) transports the oxygen dissolved in blood, carries other substances around the body, and regulates the animal's temperature.

Cut off from air

The circulatory and respiratory systems of seals face an unusual challenge. A seal is working hardest, and its body is creating the greatest demand for oxygen, when it is hunting underwater. At these times, it cannot take fresh oxygen because it must hold its breath until it surfaces. Seals, therefore, have onboard oxygen stores in their tissues, mostly in their blood. Some of the oxygen is dissolved in the blood, but most is chemically bound to hemoglobin—the protein molecule packed into red blood cells. A Weddell seal's blood has five times the oxygen storage capacity of human blood. Weight for weight, the seal has twice as much blood as a human, and the hemoglobin in its blood is 1.6 times more concentrated. Even its red blood cells are larger than those of humans, giving this seal's blood a thicker consistency.

A seal's circulatory system has many other features suited to diving. Underwater, the circulation concentrates on supplying the heart, lungs, and brain—those organs that will not tolerate a drop in oxygen level. The arteries supplying the other body parts constrict to limit blood flow. Blood going to the digestive system, muscles, skin, and flippers is reduced by 90 percent. The seal's heart now has to work less hard, and the heart rate drops from 50 and 60 beats per minute on the surface to about 15 beats per minute, greatly reducing the overall demand for oxygen.

Baggy veins

A seal's veins are wide and baggy with thin walls. They act as a reservoir for blood diverted from the seal's tissues. Instead of coursing through the seal's body, losing oxygen on the way, the blood pools in these large veins, acting as an oxygen store for the heart, lungs, and brain. The hepatic sinus, a network of baggy veins between the lobes of the liver, is huge and holds six pints (three liters) of blood in the Weddell seal. The largest vein is the Y-shape vena cava: that of

▼ **RESPIRATORY SYSTEM AND HEART**
Harbor seal
The seal's respiratory and circulatory systems can operate under pressure and for long periods without a fresh supply of air. The diaphragm is unusual because it runs at an oblique angle.

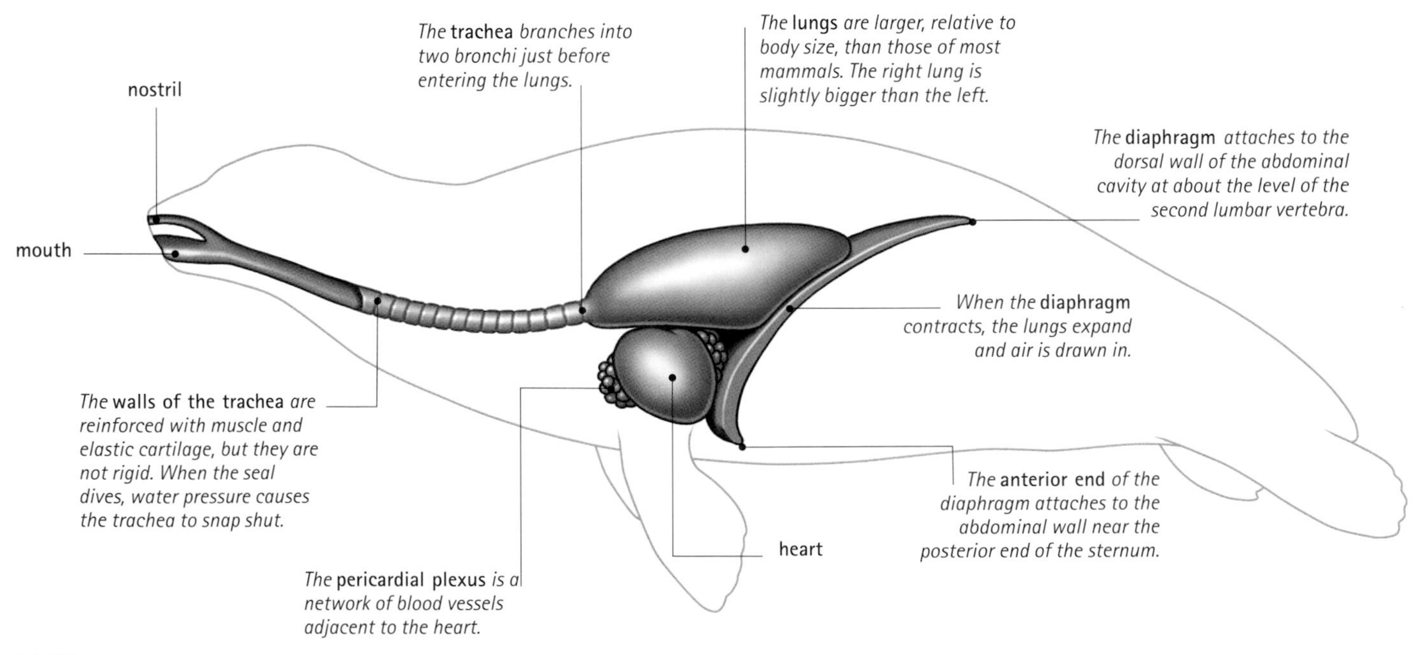

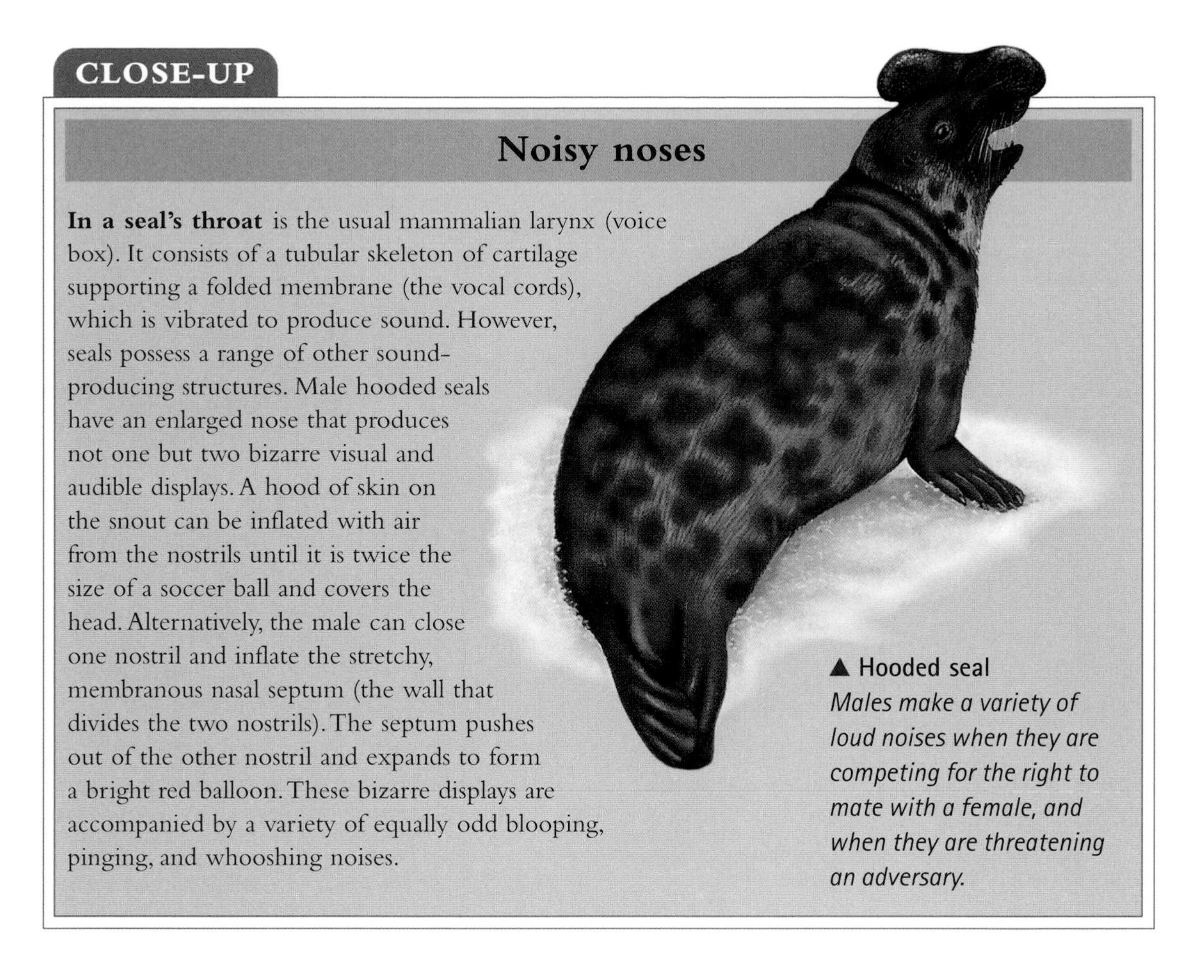

CLOSE-UP

Noisy noses

In a seal's throat is the usual mammalian larynx (voice box). It consists of a tubular skeleton of cartilage supporting a folded membrane (the vocal cords), which is vibrated to produce sound. However, seals possess a range of other sound-producing structures. Male hooded seals have an enlarged nose that produces not one but two bizarre visual and audible displays. A hood of skin on the snout can be inflated with air from the nostrils until it is twice the size of a soccer ball and covers the head. Alternatively, the male can close one nostril and inflate the stretchy, membranous nasal septum (the wall that divides the two nostrils). The septum pushes out of the other nostril and expands to form a bright red balloon. These bizarre displays are accompanied by a variety of equally odd blooping, pinging, and whooshing noises.

▲ Hooded seal
Males make a variety of loud noises when they are competing for the right to mate with a female, and when they are threatening an adversary.

the walrus is big enough for a person to pull it on like a pair of pants. The blood is held back in the veins by a muscular ring around the vena cava called the caval sphincter, which controls the return of blood to the heart.

The oxygen-starved cells and tissues continue to function using their own oxygen stores, but eventually they may begin to respire without oxygen (anaerobically). This process produces lactic acid as a by-product, which causes pain and fatigue when it builds up. Seals can tolerate a higher level of lactic acid than humans, but eventually they must return to the surface to breathe in air. Still, seals can hold their breath for very long periods. For example, a Weddell seal can last more than one hour between breaths and an elephant seal for over two hours. When the seal reaches the air, its massive veins return blood quickly to the heart, which starts pumping very quickly. Together with rapid breathing, the heart replenishes oxygen stores, and lactic acid in the tissues is broken down. Normal dives tend not to involve much anaerobic work, and the seal can usually dive again very soon.

In diving to 300 feet (90 m), the pressure on a ringed seal multiplies by 10, owing to the weight of the water above. The water pressure compresses air in the lungs, which decrease to $\frac{1}{10}$ their surface volume. The lungs of deeper-diving seals collapse completely. The construction of a seal's respiratory system copes with these pressure changes. All the airways in the lungs are reinforced with muscle and rings of elastic cartilage, but they are neither thick nor rigid and will not break. They collapse safely, then pop open again when the seal ascends.

The tiny alveoli (the chambers at the end of every airway that exchange gases with the bloodstream) are the first structures to squash flat. Next, the smallest bronchioles (narrowest airways) collapse, followed by the bronchi, and finally the trachea. This sequence ensures that air is forced from the lungs and out of the body, away from contact with the bloodstream, thus preventing gas exchange. Exchange of gas at depth is dangerous, because nitrogen in the air dissolves in the blood and can cause "the bends" when the seal ascends to the surface.

Digestive and excretory systems

CONNECTIONS

COMPARE the intestines of a seal with those of a plant-eating ***ELEPHANT***. Meat is easier to digest than plant matter, so meat-eating animals usually have shorter intestines than plant-eaters. However, seals are exceptional: like plant-eating animals, they have long intestines.

The digestive system of most mammals in the order Carnivora is fairly simple and forms a path from the mouth, through the stomach and short intestine, to the rectum. The animal matter that they eat is easy to digest—it has little of the indigestible fibrous tissue of plants. Seals are members of the Carnivora, but their digestive system differs in some significant ways.

First, seals lack the characteristic slicing carnassial teeth of other carnivores. Their teeth act only to grab and hold their prey, whether these are fish, shrimplike crustaceans, squid, or penguins. The front teeth are therefore large and pointed, and the cheek teeth are usually simple and conical, without the complex cusps and range of forms and functions seen in most mammals. There are, however, exceptions to this rule. The crabeater seal, for instance, feeds almost entirely on krill—a swarming, shrimplike crustacean. Its cheek teeth are finely divided into many lobes. It feeds by sucking krill into its mouth and then squeezing all the water out through its teeth, trapping the krill inside. The ringed seal also feeds on shrimplike animals (mysids and amphipods), but it eats many fish, such as polar cod, as well. Its teeth are not specialized in shape or function.

Seals do not slice or chew their food—they swallow it whole. The throat, or esophagus, is therefore pleated with folds that can expand to allow large food items to slip down easily. The stomach is a simple sack, similar to that of most carnivores. Unlike other carnivores, however, seals may have an enormously long small intestine. In the small intestine, digestion is completed and absorption of the liberated nutrients takes place. This is a quick and relatively simple process in the case of meat eaters, so it is unclear why elephant seals, for instance, have small intestines up to 660 feet (202 m) long. It is normal for a carnivorous mammal to have a small intestine only five or

▼ **Harbor seal**
The most notable feature of the seal's digestive tract is the very long small intestine—reaching a staggering 660 feet (202 m) long in elephant seals. Seals have a high-protein diet, and the kidneys filter out urea—a toxic by-product of protein breakdown—and excrete it in the urine.

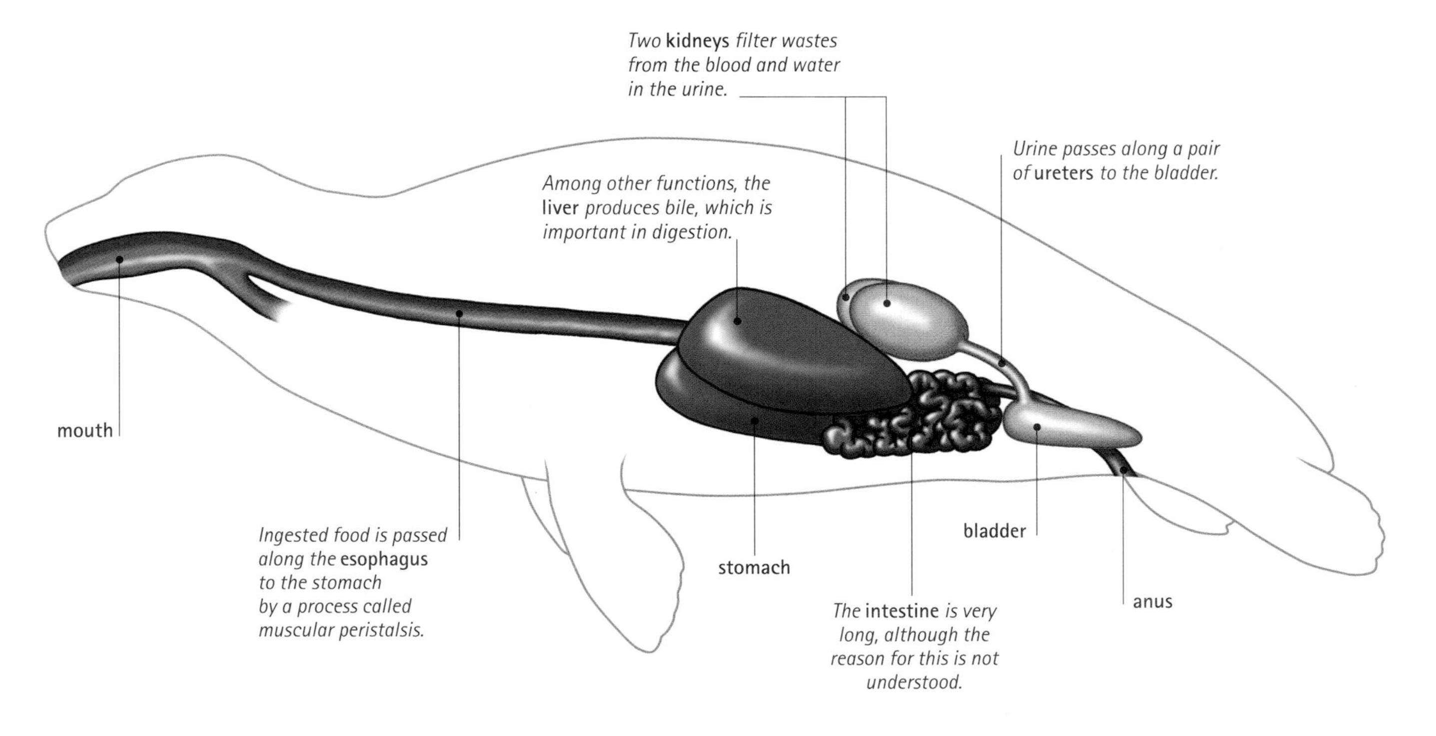

six times its body length. Also unusual is the large intestine, which has no specific adaptation for absorbing water. Most mammals recover as much water as they can from the remains of their food by absorption across the wall of the large intestine. Seals, however, do not reabsorb much water, and they pass frequent, watery feces.

Diet

The diet of seals contains plenty of fat and protein, but almost no carbohydrate. They do not receive sugars by direct breakdown of dietary carbohydrate. Their metabolism (cell chemistry) must be based on the breakdown of lipids (fats) and proteins, both of which can yield glucose. Glucose is the most common blood sugar and a vital energy source for organs such as the brain.

Another element lacking in seals' diet is freshwater. Seals have no access to freshwater, although some polar species may chew ice and snow to obtain it. However, seals get all the water they need from their prey. The breakdown of both proteins and fats in food releases "metabolic" water as a by-product. Fat yields more than twice as much water as protein does, so a thirsty seal would do well to choose to eat oily fish.

▲ *This gray seal is eating a fish. Seals eat a large range of aquatic prey, mostly fish and aquatic invertebrates. Some Antarctic species hunt and eat penguins, and sea lions sometimes eat fur seal pups.*

Water balance

The breakdown of dietary protein also produces the toxic, nitrogen-containing by-product urea. Seals must expel urea in solution in water, as urine. Owing to the loss of water in urine, seals experience a net loss of water from breaking down protein, but water from the breakdown of fat more than compensates for this. Even so, seals must constantly conserve water, especially during periods of fasting.

Production of urine is the principal role of the kidneys. They regulate levels of dissolved substances in the blood and filter out unwanted substances, to be expelled in the urine. Seals' kidneys must work intensively, because they have to remove the large quantities of salt that build up in the blood. The kidneys must also eliminate unwanted substances in concentrated form to avoid losing too much water. The kidneys are therefore large and made up of many lobes, sometimes more than 200. Each lobe functions as a miniature kidney with its own blood supply. Seals' veins branch into a complex mesh around the kidneys to keep them supplied quickly with new, salty blood to filter. Among mammals, only whales and seals have such "lobular," or "reniculate," kidneys.

IN FOCUS

Fasting

Each year, during breeding and molting, seals remain out of water. During this time they cannot eat or drink. This fast can last 90 days in the case of dominant male elephant seals in a breeding colony. Newly weaned elephant seal pups that have become fat on their mothers' milk may fast for 12 weeks before they learn to feed themselves. During these periods, seals live off their blubber. As is the case with fat in their food, breaking down the fat in their own body releases energy, the sugar glucose, and a lot of metabolic water. Even so, water must be conserved, and an elephant seal pup's urine volume reduces by 84 percent over 10 weeks of fasting, becoming extremely concentrated as the animal strives to save water.

Reproductive system

The external sex organs of seals are all tucked away in neat crevices to maintain optimum streamlining. In most male mammals, the testes hang outside the body in the scrotum, so as to keep cool. In seals, they are inside the body, underneath the skin and blubber, but not embedded in deeper tissue. It is critical for sperm manufacture, which occurs inside the testes, that the temperature is kept lower that the normal mammalian body temperature of around 99°F (37°C). The testes of seals are therefore cooled by blood flowing in from the cold hind flippers. The blood is diverted into a mesh surrounding the testes and can cool the testes by up to 7°F (4°C). This arrangement is an example of countercurrent heat exchange.

Another striking feature of a male seal's reproductive system is the large baculum, or penis bone. It is formed from the mineralized spongy tissue of the penis itself. Its function is unclear, but when the penis is not fully erect, a male seal can successfully mate with a female by using the penis bone.

The female reproductive system is of the standard Y-shape form of mammals. Each of the two ovaries is enveloped in a sac called the ovarian bursa. Eggs released from an ovary pass into a fallopian tube and down into the uterus (womb). The uterus is bicornate, and its two horns join near the cervix. Pregnancies therefore usually occur in one of the two arms of the Y-shape uterus.

Courtship and mating occur on land in the same places the females use to give birth and nurse their pups. The males and females mate, and next season's egg is fertilized as soon as the females have finished nursing this season's pup. The egg develops into a blastocyst (a hollow ball of cells), but does not implant in the uterus or develop any further for several months. The eventual implantation and development of the embryo are timed so that the pup is born almost a year later. Seals are vulnerable on land, and extending the gestation period in this way enables the seals to make just one trip to the breeding grounds a year, rather than two.

▼ **Harbor seal**
The female has two ovaries, and the uterus is bicornuate (two-horned). The male seal has internal testes, in which sperm is made, and a baculum (penis bone).

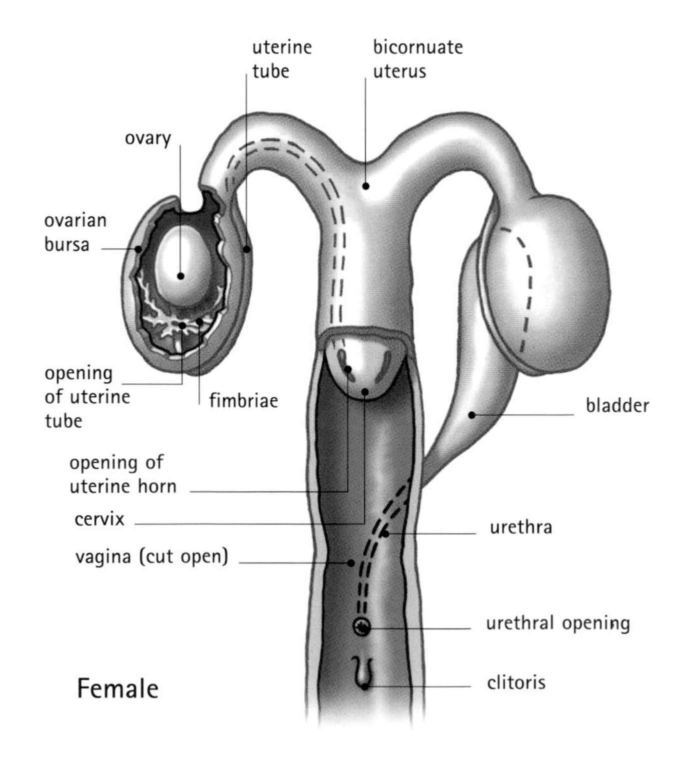

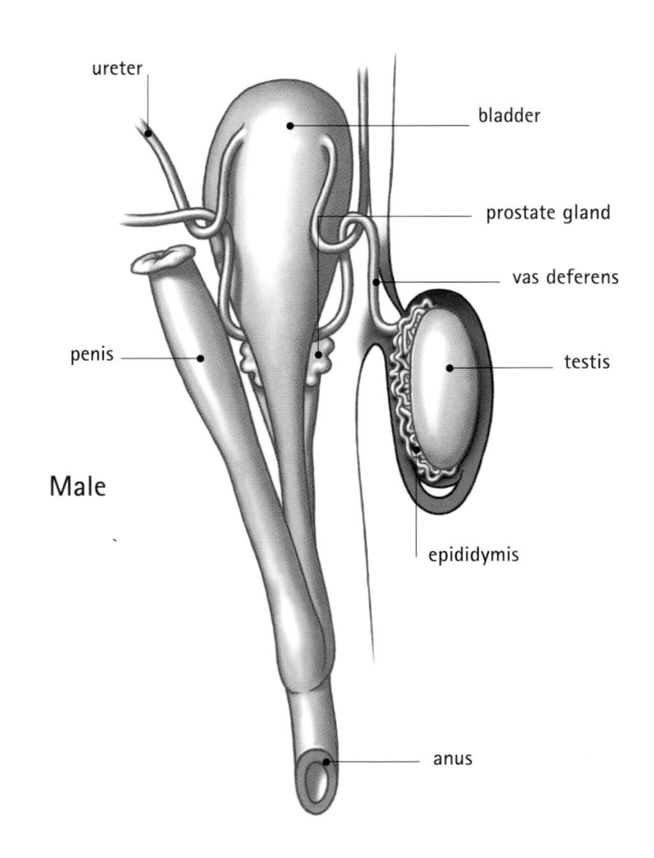

▲ *These two gray seals are mating in shallow water off the English coast. The male seal is above the female.*

CLOSE-UP

Sexual display

Different species of seal use various parts of their anatomy to attract mates or intimidate sexual rivals. The most important function of walrus tusks, for instance is thought to be sexual display. Likewise, the male elephant seal's proboscis signals his fitness as a mate. It begins growing at age two years, but it is not fully developed until the male is eight years old. When the seal is excited, the proboscis is enlarged by muscle action, by engorgement with blood, and by inflation using the lungs.

Growth spurt

Newborn pups lack thick blubber and are thus less well insulated than adults. Pups of most species have a fluffy, often white, coat (called laguno) that helps them keep warm while they work on building up a layer of fat. However, unlike the sleek adult coat, laguno is not waterproof so these fluffy pups must stay on land or on the ice until after their first molt. The precocious young of harbor and hooded seals molt before they are born, and they are able to swim almost from birth.

Seal mothers feed their young intensively on milk, in a process called lactation, from either two or four mammary glands (nipples). The nipples are positioned on either side of, and a little above and below, the mother's navel. The milk is rich and creamy, with the consistency of melted vanilla ice cream and a bland, waxy taste. In elephant seals, the milk's fat content rises from 15 to 55 percent in the first 21 days, while the water content falls from 75 to 35 percent. These changes suit the metabolic needs of the pup. At first, it needs plenty of water, but later it can make its own water by breaking down some of its newly formed blubber. Its priority then is to pile on as many pounds of blubber as it can, as quickly as possible.

Lactation is very demanding on the mother, because she is effectively transferring her blubber to her pup while taking on no food or water herself. She is expending energy six times as quickly as in resting (elephant seal mothers lose some 40 percent of their body weight during this period). The effort translates into incredible growth rate in pups. The elephant seal pup doubles its birth weight in 11 days, gaining 13 pounds (6 kg) a day. The harp seal pup puts on 5.5 pounds (2.5 kg) a day and triples its birth weight in a lactation period of only 9 days.

Snow lair

Suckling of young takes place either on land, on solid "fast" sea ice, or on pack ice. The female ringed seal uses her strong, clawed front flippers to dig a snow lair, from below, at a weak point in the fast ice. She remains close to her pup, feeding it milk for some weeks. Species that use less stable pack ice, such as the harp seal, have an extremely short, rapid form of lactation. The hooded seal is the most extreme species in this respect. It has the shortest suckling period of any mammal, completing it in an astonishing 3 to 5 days, during which the pup almost doubles in weight from 48 to 94 pounds (22–42.6 kg). Most of the weight gain is as blubber rather than growth of muscles or skeleton.

ROB HOUSTON

FURTHER READING AND RESEARCH

Bonner, Nigel W. 1989. *The Natural History of Seals*. Christopher Helm: London.

Vaughan, T. A., J. M. Ryan, and N. J. Czaplewski. 2000. *Mammalogy*. 4th edition. Saunders College Publishing: Philadelphia, PA.

Seaweed

KINGDOM: Protista PHYLA: Rhodophyta, Chlorophyta, and Phaeophyta

Seaweeds are algae—simple plantlike organisms called protists that can make their own food using energy from sunlight. Although true plants dominate the land, seaweeds colonize the ocean margins, resisting ripping currents and pounding waves. Huge brown seaweeds called kelps form dense underwater forests. Many seaweeds are edible or provide materials for industry.

Anatomy and taxonomy

Taxonomists arrange species of protists into groups based on similarities. Most protists are single-cell organisms. Seaweeds, however, are large and multicellular. Protists live in every watery environment on Earth. Many photosynthesize and so are described as "plantlike," whereas others, such as amoebas, obtain their food from other animals or plants and are described as "animal-like."

- **Algae** The algae are the plantlike protists. Most live in water or in areas that are very damp. They can be unicellular, colonial (with single cells living together, forming a larger mass), or multicellular, like the seaweeds.

- **Seaweeds** The seaweeds are marine algae. They are large multicellular organisms whose main body, or thallus, is usually visible to the human eye, and in some seaweeds can be massive. Most seaweeds grow attached to rocks. They grow from the highest reaches of the tide, where they are submerged only for a few hours a day, to the lowest depths penetrated by sunlight. A few, like sargassum, are free-floating and live in the open ocean.

Vascular plants such as the flowering plants have a well-developed network of tubes through which water, nutrients, and the products of photosynthesis flow among the roots, stems, and leaves. Seaweeds generally have no vascular system, though some of the larger seaweeds such as kelps do have a basic transport system. Seaweeds do not flower and produce seeds. Instead, they produce tiny spores, many of which can swim. This mode of reproduction, together with their biochemistry, links seaweeds with protists rather than true plants.

There are about 10,000 species of seaweeds, very few compared with the 235,000 or so species of flowering plants. The classification of seaweeds is based on their pigments

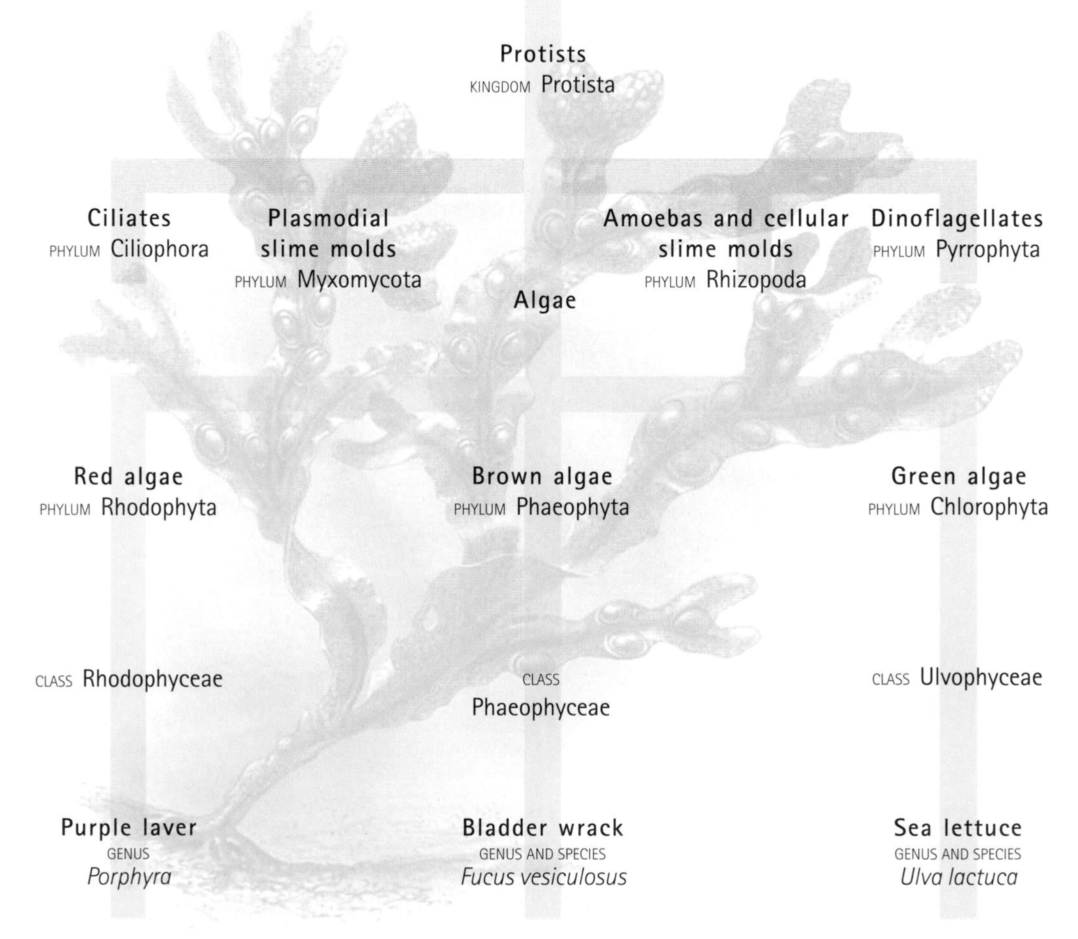

▶ *Seaweeds have been placed in several different groups but are currently included within the kingdom Protista. The taxonomy of seaweeds remains unresolved but they are generally included in three phyla of algae, according to their color: Rhodophyta (red), Chlorophyta (green), and Phaeophyta (brown).*

(colored chemicals). There are different types of the green pigment chlorophyll. All seaweeds have chlorophyll a, and some also have chlorophyll b or chlorophyll c. Seaweeds also possess other pigments: phycocyanin is bluish, phycoerythrin is reddish, carotenes are yellow-brown, and xanthophylls are brown. Seaweeds are placed within three large groups called phyla: red algae (Rhodophyta), green algae (Chlorophyta), and brown algae (Phaeophyta).

Rhodophyta The red algae form the largest group, with 6,000 species, most of which are marine. Their color can vary from pink to metallic purple. Red algae grow in a variety of shapes, from simple filaments (strings) to branched and feathery structures, thin sheets, or crusts. All red algae are multicellular organisms. Some red algae contain calcium carbonate deposits, and these species help form coral reefs.

The class Rhodophyceae is divided into two subclasses. Seaweeds in subclass Bangiophycidae reproduce by monospores. These species include the genus *Porphyra*, which is commonly known as purple laver and is eaten as flat sheets called nori in Japanese sushi bars. Seaweeds in the subclass Florideophycidae are mostly filamentous, and their cells are joined by pit connections, which are junctions containing a lens-shaped protein plug. They reproduce with a variety of spores, including monospores. Seaweeds in the genus *Gelidium*, mainly from the Pacific Ocean, are the main sources of the gelling agent agar.

Chlorophyta The green seaweeds (about 1,200 species) form only part of this large division. Most green algae live in freshwater. Green algae have the same photosynthetic pigments as vascular plants, dominated by chlorophyll a and b—hence the green color. The seaweeds within this phylum tend to be the most fragile of the seaweeds. They range from single cells to multicellular sheets and branched filaments. Most green seaweeds are in the class Ulvophyceae, which has 265 species in 35 genera. Sea lettuce has delicate broad, leaflike fronds. Seaweeds in the genus *Enteromorpha* have a strange tubular thallus, made of hollow thin-walled cylinders. The genus *Caulerpa* includes seaweeds that have a large creeping rhizome (horizontal rootlike structure) from which both rhizoids (small rootlike projections) and upright fronds arise.

▲ *Giant kelp is the largest of the Pacific kelps and can grow more than 18 inches (45 cm) in a day.*

Phaeophyta There are 1,500 species of brown algae in 250 genera. They have a variety of forms, including crusts and filaments, but most have a robust flattened thallus. Brown algae contain fucoxanthin and other gold and brown pigments, which mask the green color of chlorophyll. There is only one class of brown algae, Phaeophyceae, and it includes the kelps and the wracks. Some species of kelps are huge. For example, the giant kelp can be over 300 feet (100 m) long. Wracks are mainly intertidal species and produce mucilage (a gumlike substance), which keeps them from drying out.

FEATURED SYSTEMS

EXTERNAL ANATOMY There are red, green, and brown seaweeds. Most have a flat, leafy frond, but some have thin filaments that are feathery, tubular, or crusty. In the brown seaweeds, the frond is often supported by a rubbery midrib and held out from the anchoring holdfast by a stemlike structure called a stipe. *See pages 1128–1131.*

INTERNAL ANATOMY Seaweeds have a simple internal structure. Some are just one or two cells thick. The cells are held in a slimy matrix. Like those of plants, seaweed cells contain the pigment chlorophyll. This captures light energy for photosynthesis. *See pages 1132–1133.*

REPRODUCTIVE SYSTEM Seaweeds reproduce in a variety of ways, both sexually and asexually. Most seaweeds have swimming spores (called zoospores) at some stage in the lifecycle. *See pages 1134–1137.*

External anatomy

CONNECTIONS

COMPARE the bladder wrack with a ***DIATOM***. Both are marine algae that photosynthesize. The bladder wrack is large, multicellular, and attached to rocks, whereas diatoms are free-living, microscopic organisms. Diatoms have a silica shell, whereas bladder wrack is covered in slimy mucilage.

Seaweeds have a simple structure in comparison with that of true plants. The main difference is a lack of a vascular (water-conducting) system. Seaweeds do not need a vascular system, because they can absorb all the water, gases, and nutrients they need directly from the surrounding seawater.

Living in the sea does cause problems, though. The constant movement of currents and tides subjects the plant to extreme mechanical stress, smashing the fronds into rocks and potentially ripping the whole seaweed free. Seaweeds cope with mechanical stress by having a strong holdfast (a rootlike structure) that clings to the rock; a flexible, sometimes, rubbery stipe (stemlike structure); and fronds (the broad, leafy part of the seaweed) that bend with the waves. A thick layer of mucilage also helps protect against abrasion.

Many seaweeds live in the intertidal zone—the area of beach that is uncovered by every receding tide. These seaweeds must tolerate

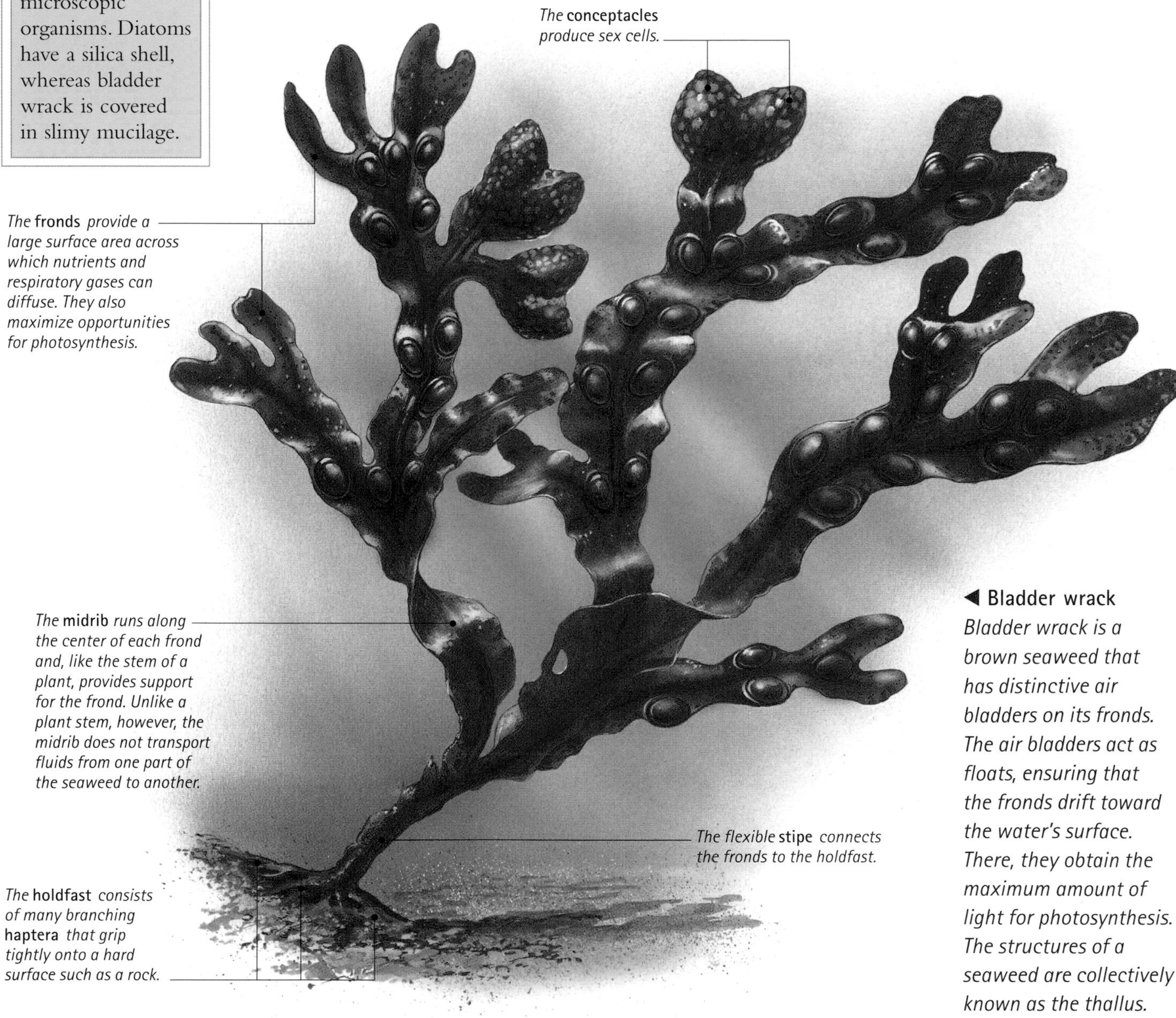

◀ **Bladder wrack**
Bladder wrack is a brown seaweed that has distinctive air bladders on its fronds. The air bladders act as floats, ensuring that the fronds drift toward the water's surface. There, they obtain the maximum amount of light for photosynthesis. The structures of a seaweed are collectively known as the thallus.

drying and the changes in temperature and salinity caused by hot sun, freezing winds, or torrential rain. Mucilage also helps prevent drying in intertidal species.

Holdfasts and stipes

The entire body of a seaweed is called the thallus. It is composed of different parts, some of which are not present in all species. The brown seaweeds, such as bladder wrack, are the most anatomically advanced of all the seaweeds. The holdfast anchors the seaweed to its substrate, whether this is rock, soft mud, the shell of a mollusk or turtle, or another seaweed. Holdfasts are usually made up of many stubby, rootlike projections called haptera. In sea lettuces, the holdfast is disk-shape, whereas in dead man's fingers it is a spongelike mass. Although holdfasts may look like roots, they do not absorb nutrients for the rest of the thallus—their purpose is solely for attachment.

The stipe is a stemlike structure that supports the leafy fronds. The structure of the stipe varies among seaweeds; it can be stiff,

▲ *The feather boa seaweed has a long branching stipe that is cylindrical at the base but flat and straplike on the upper portions. The strong holdfast has many haptera that anchor the seaweed firmly to rocks.*

◀ Sea lettuce
This is a fast-growing and short-lived green seaweed. Its fronds are just two cells thick and lose water very quickly.

frond
stipe
holdfast

◀ Kelp
Kelps are brown seaweeds that typically grow below the tide line. They have a long flexible stipe and a strong holdfast that prevents them from being torn from rocks by strong currents.

▲ Purple laver
This red seaweed has a very small holdfast and a thin, but tough, thallus that has a coating of gelatinous material.

▲ *Alaria esculenta is a brown alga with straplike fronds that grow from a short cylindrical stipe.*

flexible, solid, gas-filled, very long, short, or completely absent. The bullwhip kelp, for example, has a stipe that can grow to more than 118 feet (36 m) long, whereas the sea lettuce has no stipe at all. In the feather boa kelp, the long stipe is cylindrical at the base and branches irregularly before broadening into a flattened, bladelike frond. It is called the feather boa kelp because of the many small "feathers" and elliptical floats, or air bladders, that are attached the stipe.

EVOLUTION

Fossil seaweeds

The soft bodies of most seaweeds do not fossilize well. However, green algae of Cambrian age (543–490 million years old) and brown algae of Ordovician age (490–443 million years old) are known to have existed. Coralline seaweeds, which are red algae, are exceptional since they produce a hard calcium carbonate skeleton that preserves very well and has helped form many of our rocks.

The earliest fossils identifiable as calcium-containing algae are a now extinct group called the solenopores, from the late Cambrian period (about 490 million years ago). Fossil coralline algae that formed the groups alive today first appeared in the early Jurassic period (about 210 million years ago), when ferns and palmlike cycads covered the land.

Midribs and fronds

The midrib is a clear, firm rubberlike structure that runs along the center of each frond, from the holdfast to the tip. The midrib supports the frond, but unlike the midrib in a plant leaf, it does not contain vessels that transport fluid. In winged kelps, the midrib can be 0.4 inch (1 cm) wide and is a characteristic golden color.

The frond is the leaflike part of the thallus. It is often broad and flat, and its large surface area captures sunlight and absorbs water, nutrients, and dissolved gases that are needed for photosynthesis. In most species, the fronds are not held in a fixed position but move with the water flow—unlike the leaves of many flowering plants, which tend to be held at the best angle for capturing sunlight. In seaweeds (unlike the leaves of plants) there is no structural difference between the upper and lower surface of the fronds.

In bladder wrack, the frond branches dichotomously, that is, evenly into Y-shape branches. Wave action wears away the base of the frond so that it often disappears completely, leaving a bare midrib that resembles a stipe. Many kelps, such as those in the genus *Laminaria*, have a single bladelike frond. Tides and turbulence continually wear away the base of the blade; however, there are regions of dividing cells in this area, called meristems, so, unlike bladder wrack, the frond can regrow at the base as well as the tip. *Postelsia palmaeformis*, which grows off the Pacific coast of North America, has a frond that is divided, giving it the appearance of a miniature coconut palm.

Many intertidal seaweeds, including bladder wrack, have air bladders, or pneumatocysts. These act as floats, holding the fronds up in the water column toward the light so they can photosynthesize efficiently. The bullwhip kelp has huge air bladders, with a diameter of about 4 to 6 inches (10–15 cm), and they can be filled with up to 0.6 gallon (3 l) of gas. Sexual zones of the frond produce sex cells, or gametes. In seaweeds belonging to the genus *Fucus*, these sex cells are in swollen structures called conceptacles at the ends of the fronds.

IN FOCUS

Dividing the depths

Seaweeds have been able to colonize various niches in the sea because different species can thrive at different depths. Those high up the shore in the intertidal zone receive plenty of light but risk drying out at low tide. Others live deeper, but where the light levels are low. Seaweeds tend to grow in bands, with the brown ones near the shore, followed by the greens, and then the reds at the greatest depths.

The different light-harvesting pigments in the red, green, and brown seaweeds capture light at different wavelengths. Chlorophyll a and b absorb red light, which is available in shallow waters but absent in deeper water. Thus green seaweeds, which absorb red light, are most commonly found in the shallow intertidal zone.

The light reaching deeper waters is mainly green and blue. This light coincides with the maximum absorption of red seaweeds' pigment phycoerythrin, which absorbs green and blue light. The light energy captured by phycoerythrin is then transmitted to the chlorophyll.

Simpler seaweeds

Not all seaweeds have the relatively complex structure of the bladder wrack. Many are composed of simple sheets or filaments. Most of the red seaweeds are filamentous. Seaweeds belonging to the genus *Microcladia* (*micro* meaning "small," *cladia* meaning "branched") are very delicate, finely branched red seaweeds. *Acrosiphonia coalita* is a green alga that forms a connected mass of tangled green filaments. Each branch ends in a tiny hook that holds the mass together. The genus *Enteromorpha* contains green algae that have an unusual tubular thallus of thin-walled cylinders.

◀ *The many dividing haptera of this kelp's holdfast are clearly visible. The short stipe fans outward to produce the long straplike frond.*

Internal anatomy

CONNECTIONS

COMPARE a bladder wrack frond with the leaf of an ***APPLE* TREE**. A leaf has a complex internal structure with a vascular system, pores called stomata, and air spaces that bring water and gases together for photosynthesis. Seaweeds have a much simpler structure, since the ingredients for photosynthesis are in the water that surrounds them.

The internal anatomy of seaweeds is generally much simpler than that of plants. In many species, the thallus is just one or two cells thick. In the genus *Monostroma*, growth begins as a filament, but this structure develops into a flat, leaflike thallus just one cell thick, attached to rocks by a holdfast. A sea lettuce frond has a similar appearance but is two cells thick. *Nemalion multifidum* is a simple, branched red alga composed of chains of cells enveloped in mucilage. The function of the mucilage is not fully understood, but it probably protects against attacks by pathogens.

CLOSE-UP

Large chloroplasts

The double-layered cells of a sea lettuce each have a single large chloroplast, which takes up almost half the cell volume. The sea lettuce was the first organism in which scientists demonstrated that during cell division the chloroplast divides at the same time as the rest of the cell.

Photosynthesis takes place in the meristoderm.

cortex

medulla

The cells that make up the cortex *do not possess chloroplasts.*

The cells in the meristoderm *layer can divide.*

▲ CROSS SECTION THROUGH FROND
Bladder wrack
Photosynthesis and cell division occur in the meristoderm. The medulla provides support.

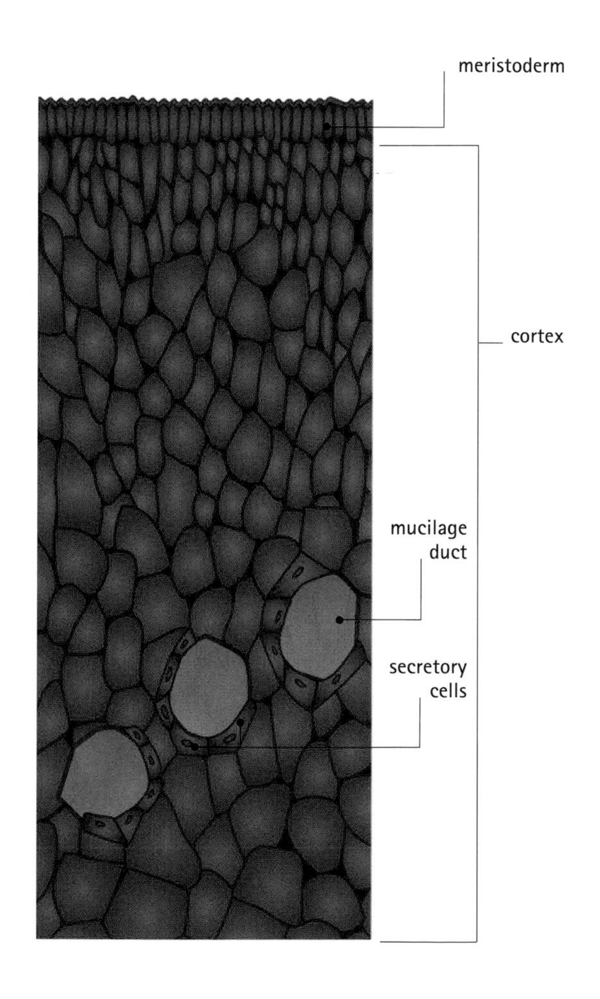

▲ CROSS SECTION THROUGH FROND
Giant kelp
Mucilage is produced in small cells that run along canals called secretory ducts.

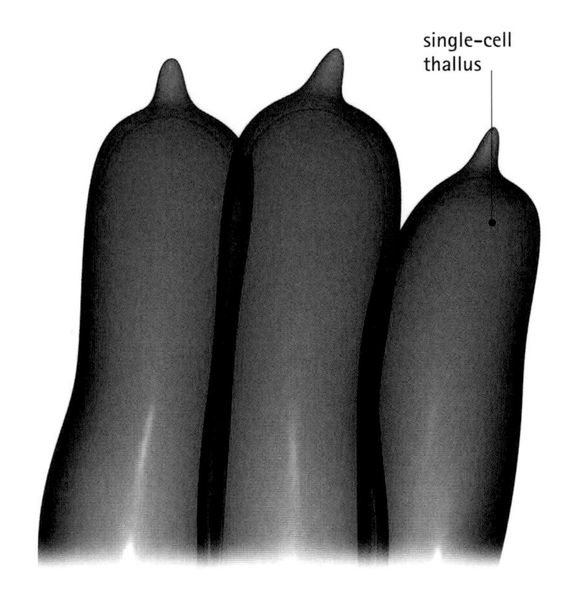

▲ UTRICLES
Dead man's fingers
The spongy thallus of a dead man's fingers obtains its green color from a green layer of vessicles called utricles that contain chloroplasts.

COMPARATIVE ANATOMY

Giant cells

Most seaweeds are multicellular, with each cell well defined by a boxlike cell wall. A few seaweeds, though, have an unusual way of growing. Instead of boxing off compartments with cell walls, the whole thallus is one giant cell with many nuclei. Seaweeds like this are called coenocytic. Dead man's fingers is one example. This seaweed is dark green, with soft, cylindrical branches that can grow to 12 inches (30 cm) long. Another coenocytic seaweed is the mosslike *Bryopsis corticulans*, which forms dense patches on rocks in areas exposed to heavy surf.

▼ *Dead man's fingers is a green seaweed that grows on intertidal and upper subtidal zones of temperate waters. There, its single-cell thallus clings to the tops and sides of rocks.*

In the brown seaweeds, there is more internal specialization than in most of the red and green seaweeds. A *Laminaria* stipe, for example, has three distinct zones in cross section. The outer layer is covered with thick mucilage. This layer is called a meristoderm because it is meristematic: that is, the cells can divide. The cells of the meristoderm have chloroplasts (organelles that contain photosynthetic pigments). Below this layer is a chloroplast-free zone of paler, long cells, forming the cortex. At the center of the stipe is a slimy matrix (medulla) that contains intertwining, branched filaments that give strength to the structure. The innermost part of the cortex and medulla also contains columns of long cells broadened at each end, which are referred to as "trumpet hyphae." These look very much like phloem cells in higher plants, and, like phloem, they are used to transport sugary products of photosynthesis to the lower portions of the thallus. These may be many feet below the fronds.

Seaweed cells contain cytoplasm, a nucleus, one or more chloroplasts, and other typical cell organelles, surrounded by a membrane and cell wall. The cell wall is a boxlike structure that forms outside the cell membrane. Plant cells also have cell walls, but animal cells do not. The cell walls of seaweed are usually thick and offer protection against abrasion and help keep intertidal seaweeds from drying out. This layered cell wall is usually composed of cellulose (a long, fibrous molecule), hemicellulose (a binding molecule), and slimy polysaccharides (chemicals made of chains of sugar units). These polysaccharides, called alginates, are used as thickeners and emulsifiers in the food, paint, and cosmetics industries.

The long, thin cells that make up the thallus of many calcium-containing red algae are joined by pit connections. Each pit is a ring with a lens-shape protein plug held in the walls by grooves. Pits can form both at the ends of the filamentous cells and in the side walls.

Reproductive system

CONNECTIONS

COMPARE the gametangia of the bladder wrack with the ovary and anther of a ***MARSH GRASS***. Both produce sex cells, but in the marsh grass the male sex cells drift in the wind, whereas in the bladder wrack the male sex cells swim to the female sex cells.

Seaweeds reproduce in a variety of ways. The simplest form is asexual reproduction, which produces genetically identical clones of the parent. Asexual reproduction can be vegetative propagation, in which fragments of the parent thallus break off and grow into a new individual. Sargassum reproduces in this way. Alternatively, cells in the thallus divide to produce asexual spores that are released into the water, settle, and grow into new seaweeds, as in the monospores of *Porphyra*. Spores released into the water can be carried for thousands of miles and are responsible for the wide distribution of many species. In many seaweeds, the spores can swim, using tail-like flagella. These mobile spores are called zoospores because of their animal-like ability to swim through the water.

Sexual reproduction "shuffles" the genes on the chromosomes so the offspring are slightly different from the parents. This variety increases the chance that some will survive in the unpredictable environment of the ocean. In sexual reproduction, sex cells, or gametes, are produced by meiosis, a form of cell division that halves the chromosome number in a cell. These cells with only one set of chromosomes are called haploid. When two haploid gametes fuse to form a zygote, the full complement of chromosomes is restored. Cells with a full set of chromosomes are called diploid.

Life cycles in algae

The alternation between haploid and diploid states that occurs in sexual reproduction allows for a variety of scenarios. Seaweeds are a diverse group that demonstrates at least three variations in ways of moving between haploid and diploid states. In the simplest case, the diploid state is represented only by fusion of

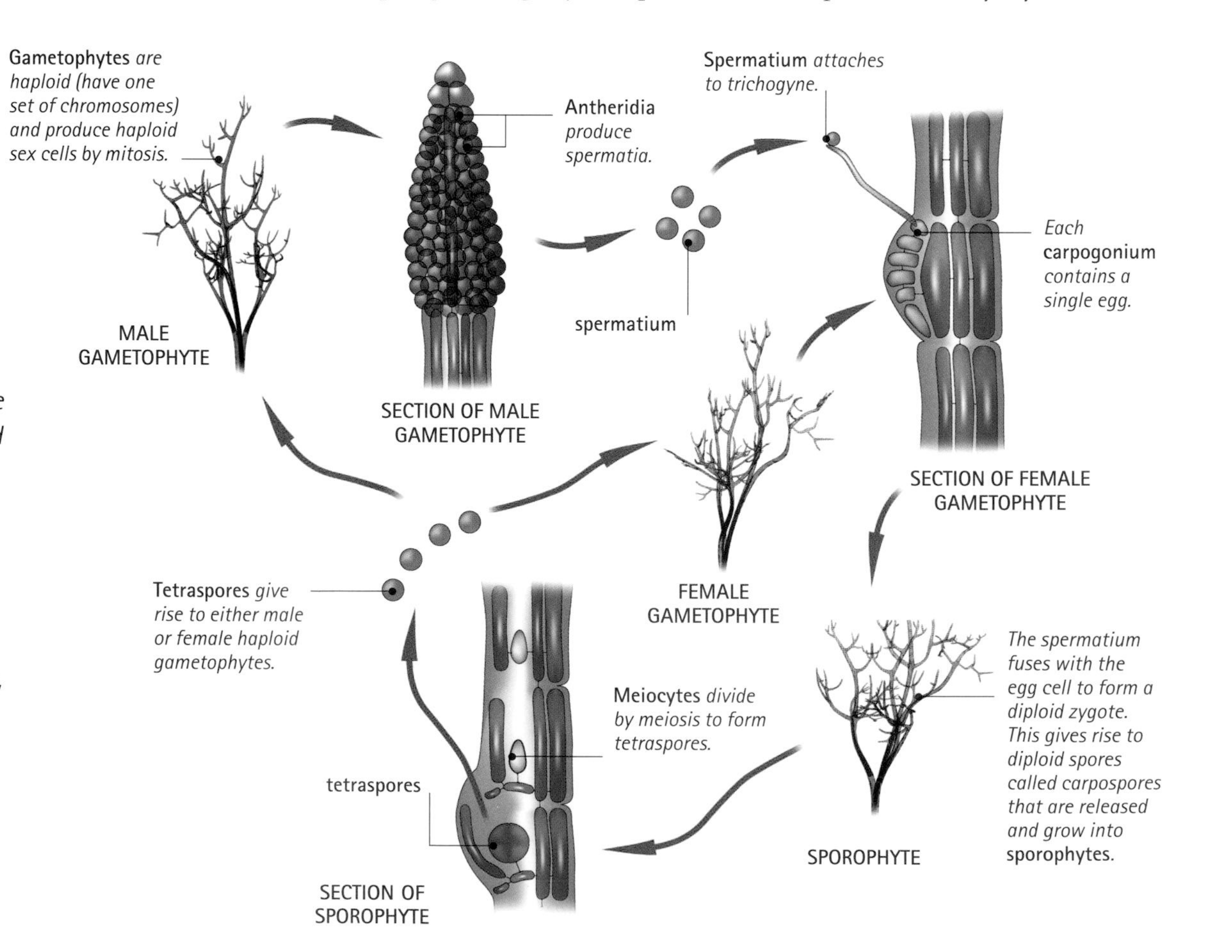

▶ LIFE CYCLE
Polysiphonia
In this red alga, separate male and female haploid gametophytes produce spermatia and eggs by mitosis. Once an egg is fertilized, the resulting zygote gives rise to carpospores that grow into diploid sporophytes, which produce haploid tetraspores by meiosis. The tetraspores grow into either male or female gametophytes, and the life cycle continues.

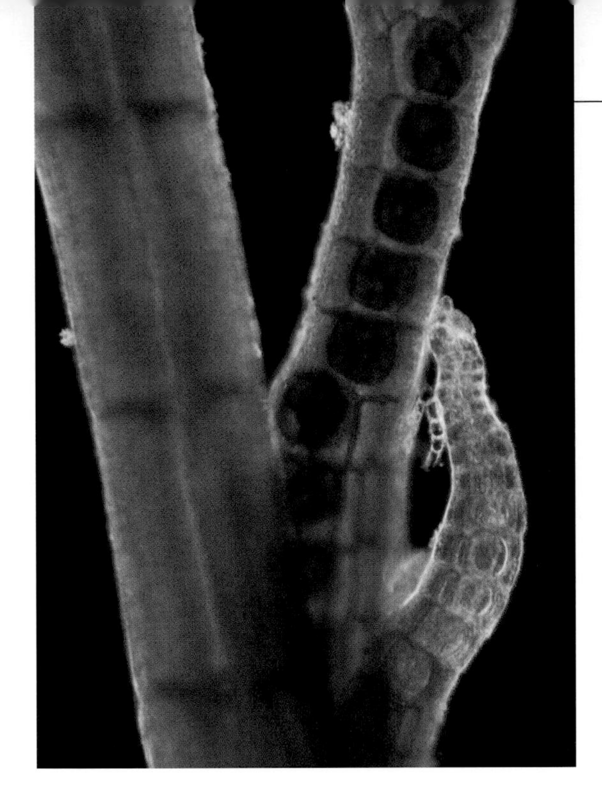

▲ *This image shows the sporophyte of the red alga* Polysiphonia. *The opaque, round structures in some of the cells are tetraspores. When released, these haploid spores (spores with only one set of chromosomes) grow into the haploid generation.*

similar-looking gametes to produce a zygote. This zygote usually forms a tough resting spore. When germination occurs, the zygote immediately undergoes meiosis to produce haploid algal cells. The simpler algae, including filamentous green alga and diatoms, undergo this type of reproduction.

At the next level of complexity, the diploid zygote delays meiosis and grows, producing a diploid sporophytic (spore-producing) phase. This eventually produces reproductive bodies, almost always zoospores, by a process involving meiosis. The spores grow into a haploid body that produces gametes (a gametophyte). In sea lettuce, the haploid and diploid forms look the same (like floppy lettuce), but in many species the two phases can look very different and have often been mistaken for different species.

In the third scenario, the diploid condition predominates, as is the case in most animals and higher plants. Bladder wrack undergoes this type of reproduction. The algal thallus is called a sporophyte because it produces spores. The haploid stage is represented only by gametes, and meiosis occurs during gametogenesis (the process of producing gametes).

IN FOCUS

Polarity

When a zygote (a fertilized cell) of *Fucus* settles on a rock or another substrate, it divides into two. The direction of this first cell division plane is controlled by a basic "sense of direction" in the cell, which responds to gradients in light, temperature, or acidity. Once the cell detects a gradient—for example in light direction—movement of calcium ions across the cell sets up an electrical gradient. The cell divides across this gradient, so the two offspring cells are at different ends. The cell that is closest to the light develops into the stipe and frond, whereas the lower cell develops into the holdfast.

Red algae reproduction

In red seaweeds, the spores, whether produced sexually or asexually, never have a flagellum for swimming (they are nonmobile). Instead they drift until they settle, at which point they grow into a new thallus. The thallus of *Porphyra* is a sheet normally just one cell thick. This seaweed can reproduce sexually and asexually. Asexual

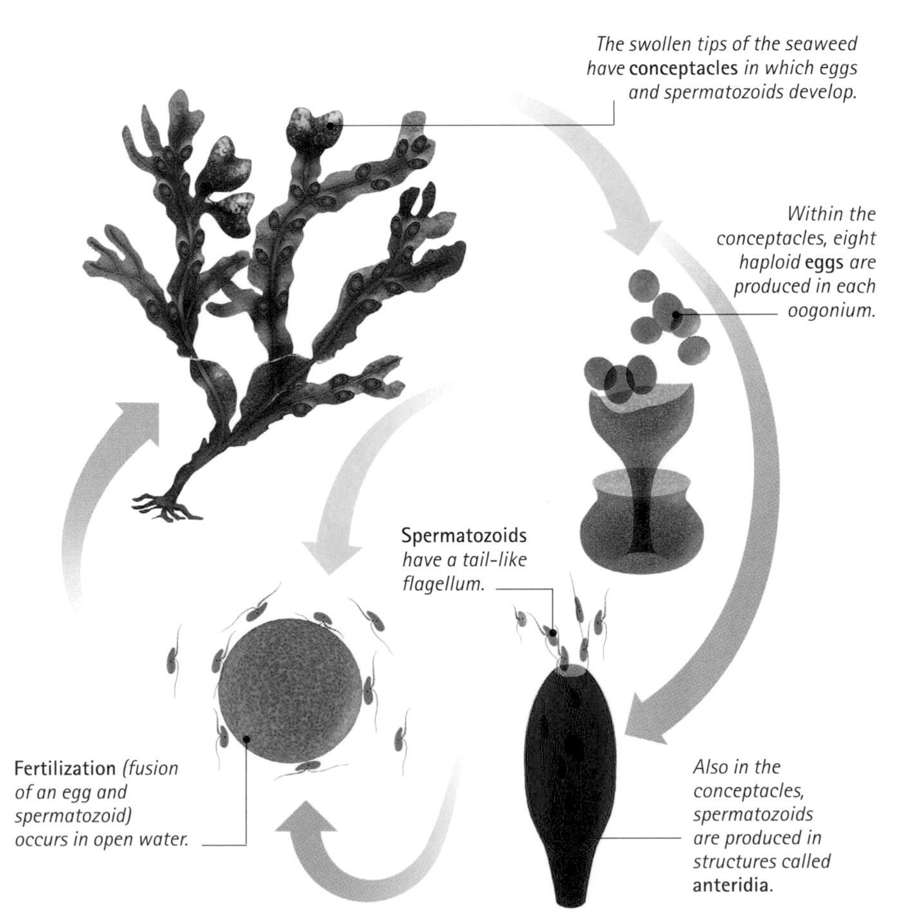

▼ REPRODUCTION
Bladder wrack
This seaweed has a life cycle similar to a flowering plant in that there is no free-living haploid gametophyte stage.

reproduction is by monospores, which are produced simply by division of the vegetative cells of the frond, to produce clusters of spores, in fours. These are released from the surface or margins of the sheet into the water.

When *Porphyra* reproduces sexually, cells in the thallus divide many times, with each cell producing 64 or 128 tiny cells called spermatia (designated as male cells because of their size). Carpogonia (cells housing the female structures) develop from cells on the thallus. Each carpogonium contains a single egg. The spermatia drift to the carpogonia and become trapped in the mucilage that surrounds them. A chemical signal triggers the spermatium to push out a tubular cell extension. This extension pierces the egg cell wall and membrane, allowing the body of the spermatium to be drawn into and fuse with the female cell.

Fusion of the spermatium and female cell results in fertilization and the production of a zygote, which divides and releases several carpospores (spores from a carpogonium). These spores settle, often on an oyster or mussel shell, and grow into a small filamentous alga. Before this alga was recognized as being a phase in the life cycle of *Porphyra*, it was called *Conchocelis*. This little filamentous alga can also reproduce in two ways: either vegetatively by monospores, or by another type of spore (called a conchospore) that regenerates the familiar sheetlike gametophyte.

Sea lettuce: A double life

In sea lettuce, both generations of the life cycle are large and leafy. When sea lettuce undergoes sexual reproduction, haploid thallus cells divide, producing up to 32 zoospores. These are tiny, and all are produced inside the rigid

▼ *The swollen yellowish structures of this bladder wrack contain conceptacles, which are the parts of the seaweed that produce sex cells, or gametes. Not all seaweeds produce sex cells in this way.*

COMPARATIVE ANATOMY

Alternation of generations

Switching between a haploid (single set of chromosomes) and diploid (double set) condition is called "alternation of generations." In many seaweeds, both haploid and diploid stages can form an obvious, identifiable thallus. In ferns, by contrast, the plant we see is the diploid sporophyte, which produces spores. These germinate and grow into tiny haploid gametophytes where fertilization takes place. In flowering plants, animals, and many seaweeds, there is no free-living haploid stage.

cell wall of the parent cell. Zoospores have a flagellum at one end and so can swim. All the zoospores are the same size, so they cannot be differentiated into male and female. Instead they differentiate into mating types. The two mating strains are produced in equal numbers. (By contrast, in *Porphyra* hundreds of tiny male cells are produced per single egg cell.) Zoospores from the same thallus cannot mate with each other—they have to cross-fertilize. After fertilization, the zygote grows into a diploid thallus that looks like the haploid parent plant. More zoospores are produced, this time by meiosis. These settle to produce another haploid thallus. Most of the vegetative cells on a thallus produce zoospores, and after they have been shed, the parent thallus remains as a bleached sheet of empty cells.

Plantlike sex life

In bladder wrack, the tips of the fronds are swollen with tiny blisterlike structures dotting the surface; these are called conceptacles. Some of them contain male gametangia (cells that make gametes), and others female gametangia. The conceptacles are lined with sterile hairs called paraphyses. The gamete-forming cells undergo meiosis to produce haploid gametes. Both male and female gametes are shed into the water. The egg cells drift passively, but the male gametes (spermatozoids) have a flagellum and swim toward the eggs.

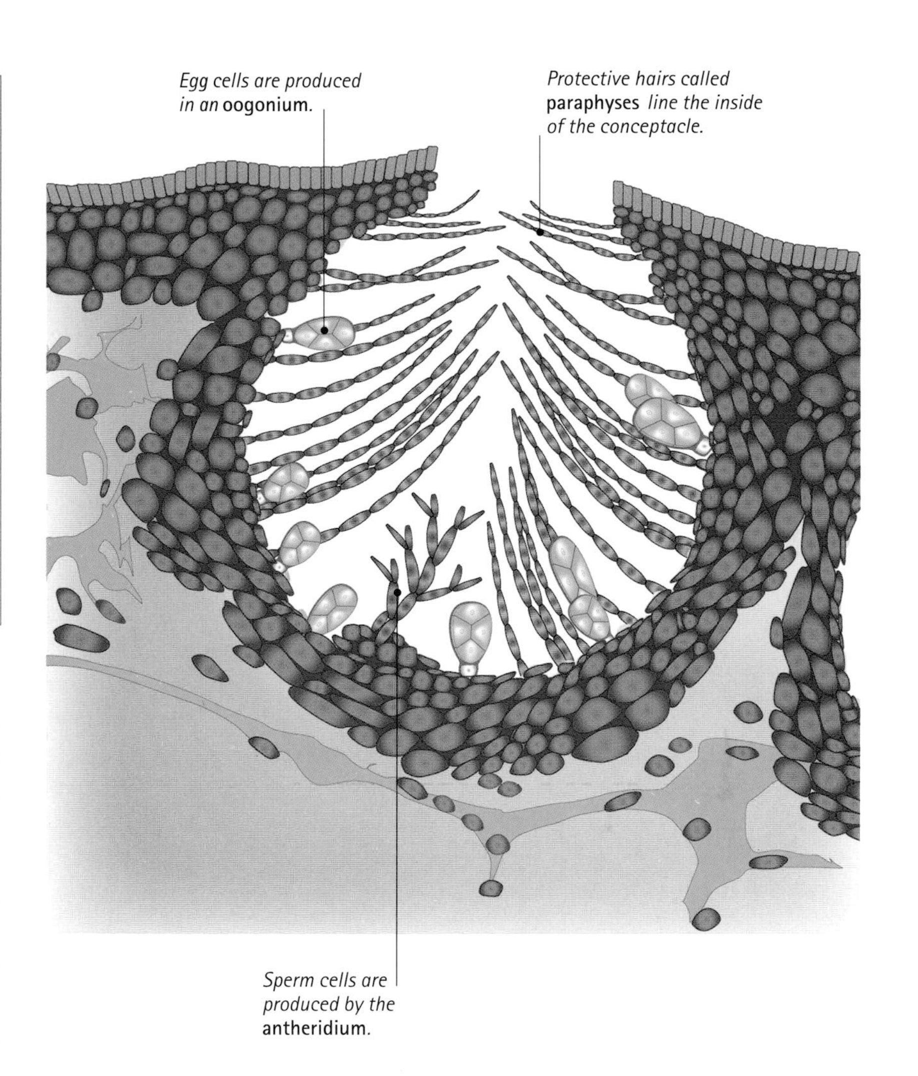

▲ SECTION THROUGH A CONCEPTACLE
Bladder wrack
To reproduce sexually, bladder wrack produces sex cells in cavities called conceptacles at the swollen ends of some of its blades. The eggs form in oogonia, and the sperm cells in branched structures called antheria.

After fertilization, the zygote continues to drift, then eventually settles and becomes anchored by mucilage. It grows into the familiar seaweed sporophyte. There is no free-growing gametophyte stage. This arrangement is similar to flowering plants, in which the gametophytes are the pollen and the ovule. Bladder wrack does not have specialized asexual reproduction.

ERICA BOWER

FURTHER READING AND RESEARCH

Bell, P. R., and A. R. Hemsley. 2005. *Green Plants: Their Origin and Diversity*. Cambridge University Press: New York.

Van den Hoek, C., D. Mann, and H. M. Jahns. 1996. *Algae: An Introduction to Phycology*. Cambridge University Press: New York.

Rocky intertidal shores: www2.mcdaniel.edu/Biology/wildamerica/rockyforweb/rockyshores.html

Sequoia

KINGDOM: Plantae ORDER: Coniferales FAMILY: Taxodiaceae
GENUS: *Sequoiadendron*

Conifer trees hold many records. The giant sequoia is the world's most massive tree. Other conifers are the tallest and oldest trees. Vast conifer forests once covered much of Earth. They still make up 30 percent of our forests and are our main source of timber.

Anatomy and taxonomy

Botanists arrange species of plants into groups that are based on similarities. However, they do not always agree on which characteristics should be used to judge similarity, so there are many systems in use that differ slightly from one another. Sequoias are classified in the family Taxodiaceae.

• **Plants** All true plants are multicellular. One of the key characteristics of most plants is that they are green. Unlike animals and fungi, most plants can make their own food using the energy from sunlight to turn simple chemicals in the air, soil, and water into carbohydrates (sugars). This process is called photosynthesis (from *photo,* meaning "light," and *synthesis*, "manufacture") and is carried out by the green molecule chlorophyll.

• **Vascular plants** The simplest plants (mosses and liverworts) have no specialized tissue for transporting water, so they are small and need to live in damp places. When plants evolved water-conducting (vascular) tissues, they could grow larger and move into drier habitats. The first vascular land plants evolved about 420 million years ago (Silurian period) and included ferns, club mosses, and horsetails. These plants reproduce using spores rather than seeds. They dominated Earth until the seed-bearing plants evolved during the late Devonian period, 375 million to 360 million years ago.

• **Seed-bearing plants** There are two major groups of seed-bearing plants: gymnosperms and angiosperms. Angiosperms bear their ovules inside an ovary, which ripens into a fruit containing the seeds. These are flowering plants and include most of our food plants, such as cereals and potatoes; broad-leaved trees such as apple trees; and ornamental flowers.

• **Gymnosperms** Gymnosperm means "naked seed": seeds are not enclosed in an ovary. There are four major gymnosperm groups, or phyla: Ginkgophyta, Gnetophyta, Cycadophyta, and Coniferophyta.

The ginkgo, or maidenhair tree, is an ancient species from China. Fossil ginkgos that are 200 million years old and lived before the dinosaurs looked identical to modern gingkos. Cycads are also an ancient group of plants. They look like woody ferns, but produce cones with seeds in them—true ferns produce spores. Gnetophytes include the unusual-looking *Welwitschia* of southwestern Africa.

• **Conifers** Almost all conifers are evergreen trees. Many survive in extremely hostile environments, from parched deserts to frozen mountains. The typical conifer has either scales or needlelike leaves, and cones rather than flowers and fruits. Most conifers produce sticky resin and have secondary wood composed of tracheids (long water-conducting cells with pits).

There are about 600 species of conifers in 70 genera, grouped in nine families: Araucariaceae, Cephalotaxaceae, Cupressaceae, Phyllocladaceae, Pinaceae, Podocarpaceae, Sciadopityaceae, Taxaceae, and Taxodiaceae.

▼ *This family tree shows that the giant sequoia is a conifer and a member of the family Taxodiaceae, which also includes the California redwood. Only living species are shown below.*

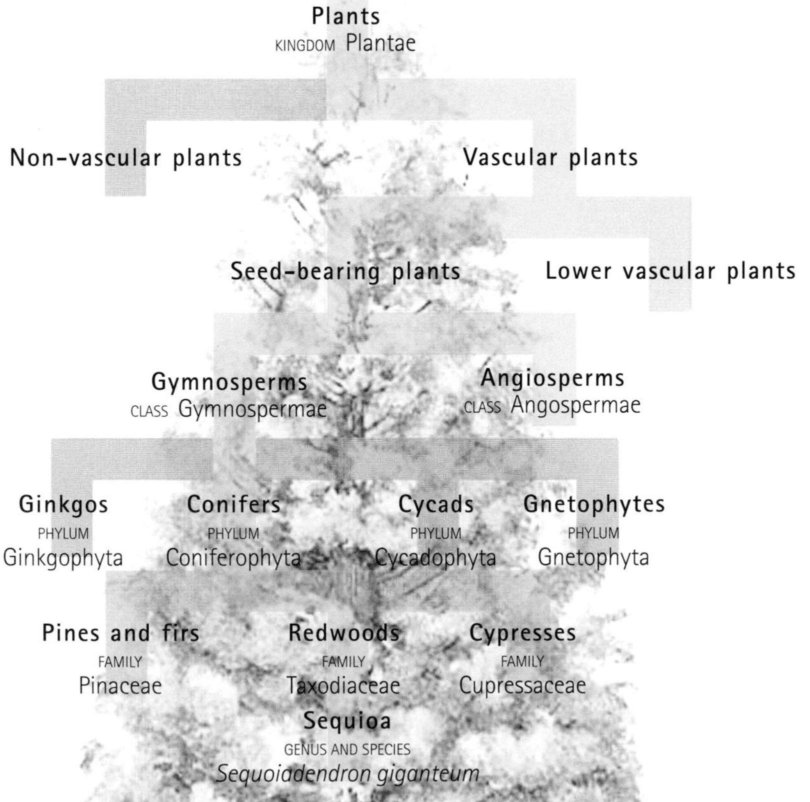

The Araucariaceae is an ancient family, with only three living genera, *Araucaria*, *Agathis*, and *Wollemia*. The monkey puzzle tree (*Araucaria*) from Chile is one of the most distinctive conifers, with a bare trunk and a crown of branches covered by spirals of leathery, triangular leaves. The Wollemi pine is an ancient species that was discovered as recently as 1994 in a remote part of eastern Australia.

Plants in the family Cupressaceae have needlelike leaves when young, but these usually change to scales as the tree matures. The cones are woody, leathery, or berrylike. This family includes the Lawson cypress, junipers, and the Monterey cypress. The popular garden conifer Leylandii cypress is a hybrid of the Monterey and Nootka cypresses.

The family Pinaceae contains some of the most familiar conifers: pines, cedars, spruces, and firs. The leaves of these trees are long and thin, often needlelike. Many are timber trees. Bristlecone pines live in the deserts of California, Nevada, and Utah and are the world's longest-lived trees, many more than 5,000 years old. The family Taxaceae contains the yews; they can also live for thousands of years.

• **Taxodiaceae** The family Taxodiaceae has 16 species in 10 genera, living in North America, eastern Asia, and Tasmania. As in most conifers, the leaves are long and thin, often needlelike. The California redwood, or coastal redwood, is the tallest living thing. One tree was measured at 345 feet (112 m). The record for the largest circumference of any tree is held by a Montezuma cypress, from Mexico: 118 feet (35.8 m).

• ***Sequoiadendron*** The giant sequoia (also called "big tree," wellingtonia, and Sierra redwood) is the only species in the genus *Sequoiadendron*. It grows on the western slopes of the Sierra Nevada, California, and is the most massive tree on Earth. One tree, the General Sherman, is 270 feet (84 m) tall, has a trunk 33 feet (10 m) in diameter and 102 feet (31 m) in circumference, and has an estimated weight of 6,600 tons (6,000 tonnes). The oldest giant sequoia is estimated to be about 3,500 years old.

Fossil evidence shows that the giant sequoia has changed little in the last 20 to 30 million years and once flourished around the world. Since 1850, the timber trade has reduced the Californian forests to just 72 groves of giant sequoias.

▲ *The towering giant sequoia is native to the western slopes of the Sierra Nevada, California. Although it does not grow as tall as its close relative the California redwood, the giant sequoia has a thicker trunk and is the world's most massive tree.*

EXTERNAL ANATOMY The giant sequoia, like most conifers, has a strong leading shoot and shorter side branches, giving the tree a tall, conical shape. Thick, spongy bark protects the tree from forest fires. Scalelike leaves minimize water loss. *See pages 1140–1143.*

INTERNAL ANATOMY The wood of conifers consists mainly of water-conducting tubes called tracheids. In the center of the giant sequoia trunk, the heartwood is stained red. Leaves and stems of conifers usually contain resin ducts. *See pages 1144–1145.*

REPRODUCTIVE SYSTEM The giant sequoia and other conifers have cones rather than flowers. They are wind-pollinated. Their seeds are naked rather than being enclosed in an ovary. Because there is no ovary wall, conifers cannot form fruits. *See pages 1146–1149.*

External anatomy

CONNECTIONS

COMPARE the shape of an ***APPLE TREE*** and other hardwood trees with that of a giant sequoia. In an apple tree, no single shoot grows faster than any other, and this trait results in a tree with a rounded shape with many large branches. In a giant sequoia, the leading tip at the top of the trunk grows faster than any other shoot, giving the conical shape typical of conifers.

Conifers are usually easy to recognize from a distance by their shape. The leading shoot at the top of the trunk is usually dominant and grows faster than the side branches. This arrangement results in the typical tall, pointed Christmas-tree shape. Not all conifer trees are conical, though. The Montezuma pine, which is native to the mountains of Mexico and Guatemala, has a broad, spreading shape that from a distance looks much more like that of a deciduous tree. The Chinese plum yew also has a broad, almost shrublike shape, and the Rocky Mountain juniper is tall and columnar. The shape of a tree can also be dictated by its environment. On an icy, wind-blown tundra, conifers grow low, hugging the ground.

▼ *The branchlets of a giant sequoia have green, overlapping scalelike leaves that produce food for the tree by photosynthesis. The cones may stay green and closed for 20 or so years before drying out and opening to release winged seeds.*

CLOSE-UP

Forever green?

The term *evergreen* is used for trees that have leaves throughout the year, unlike deciduous trees that shed leaves for a season. Most conifers are evergreen. Their leaves do eventually drop off, but there is overlap, so the tree is never bare. In some species, the overlap is very short, and the previous year's needles are shed as soon as the new ones grow. The needles of the bristlecone pine, which lives in harsh environments, can live for 20 to 30 years. Adding new leaves takes energy and nutrients, so long-lived leaves enable the plant to carry on photosynthesizing even when times are tough. In a sequoia, the leaves do not drop off the tree individually. When the tree does lose leaves, the whole branchlet is shed. However, a few conifers, such as larch and dawn redwood, shed their needles in the fall and remain bare until the following spring.

▲ *The monkey puzzle tree is an ancient conifer native to the slopes of the southern Andes mountains. It has distinctive leathery, triangular scales and grows up to 100 feet (30 m) tall.*

▼ **Giant sequoia**

The giant sequoia has a conical shape characteristic of most conifers. Its most impressive feature is its massive trunk. The trunk transports water and nutrients from the roots to the scalelike leaves on the branches. In addition, the branches and trunk convey sugars made by photosynthesis in the leaves to all living parts of the tree.

The **branches** *bear scalelike leaves and male and female cones.*

The massive **trunk** *has reddish-brown bark, which is fibrous, spongy, and deeply ridged.*

The dense network of **roots** *is shallow but spreads over a wide area. The roots anchor the tree and extract water from the soil.*

A supporting trunk

The trunk is the essential feature of a tree, distinguishing it from a shrub, which has several stems rather than one main trunk. The trunk has two main functions: it supports the branches and leaves, and it transports water, soil nutrients, and other chemicals between the roots and the leaves. The outside of the trunk and branches is covered with a layer of bark. Conifers have many different types of barks: fibrous, wrinkled, ridged, scaly, or smooth. In the giant sequoia, the reddish bark is very fibrous and spongy, and ridged, in patterns that can be vertical, spiral, or netlike.

▲ BARK

Giant sequoia

The red, spongy bark of the giant sequoia can grow more than 1 foot (30 cm) thick. The bark is made of fibers, which break off in pieces, littering the base of the tree.

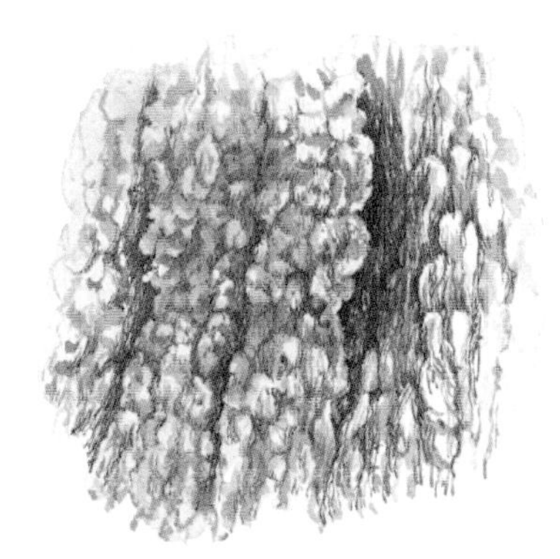

▲ BARK

Larch

The larch—a conifer in the pine family—has a grayish-brown outer bark with loose plates that reveal a reddish inner bark.

COMPARATIVE ANATOMY

California redwoods and giant sequoias

The California redwood and the giant sequoia were once both classified in the genus *Sequoia*, but in 1939 the giant sequoia was given its own genus, *Sequoiadendron*, to reflect differences between the species. Both have reddish-brown bark and grow to great heights. However, the California redwood will regrow from cut stumps and produce suckers (new shoots) from roots, as the giant sequoia never does. In addition, the giant sequoia has only 22 chromosomes, whereas the California redwood has 66 chromosomes. Botanists have speculated that redwoods originated from hybridization between a living species of conifer called a metasequoia and another now extinct conifer. This hybridization is thought to have occured during the late Mesozoic era (70 million to 120 million years ago).

Branches and leaves

The branches hold the leaves away from the trunk, so the tree is able to gather sunlight from a wide area. In most conifers, branches grow out from the trunk in regular whorls (rings around the stem), with bare sections lacking branches between the whorls. This regular pattern is most obvious in the monkey puzzle tree but can be easily seen in most other conifers.

Leaves are a tree's food factory, where carbon dioxide from the air and water from the soil are converted into sugar using energy harnessed from sunlight. Conifers tend to have linear, needlelike or scalelike leaves, rather than the broad, flat leaves of most angiosperms. Needles of pine, fir, spruce, cedar, and larch trees are long, thin, and sharp. In some species, the needles emerge in clusters, whereas others, such as Californian redwood, have long, thin, bladelike leaves in pairs along a central axis.

Scalelike leaves are found in many conifer families. The monkey puzzle tree has large scales that are flattened, leathery, and triangular. The giant sequoia has spreading scales that overlap each other in spirals along the twig. Each scale has a sharp pointed tip. The Italian cypress has overlapping scales that completely clothe the twig. The kauri is unusual among conifers in having flattened, fan-shape leaves.

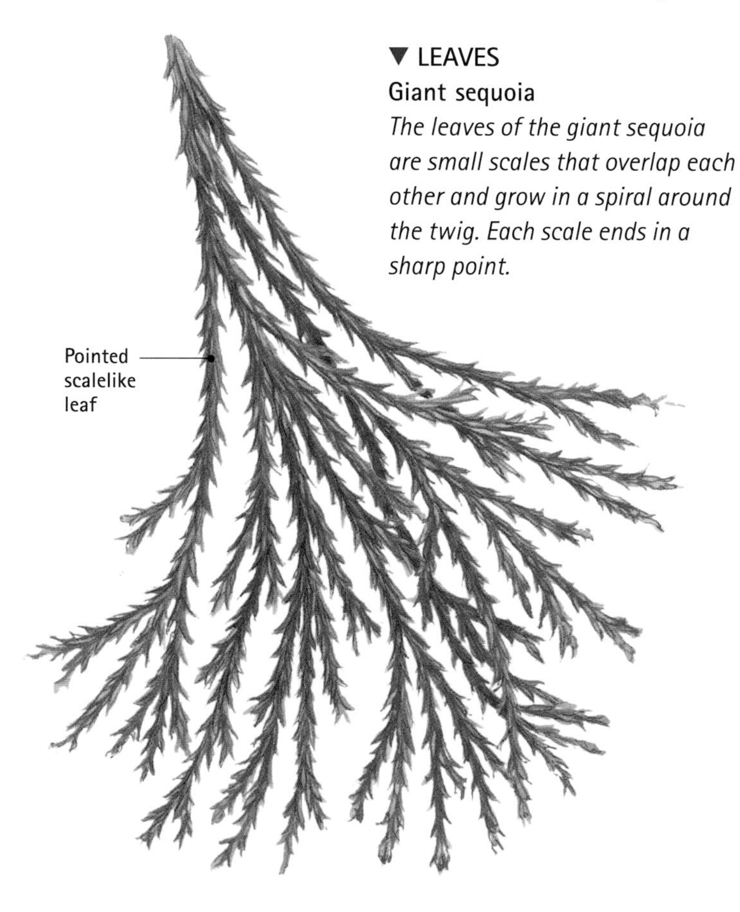

▼ LEAVES
Giant sequoia
The leaves of the giant sequoia are small scales that overlap each other and grow in a spiral around the twig. Each scale ends in a sharp point.

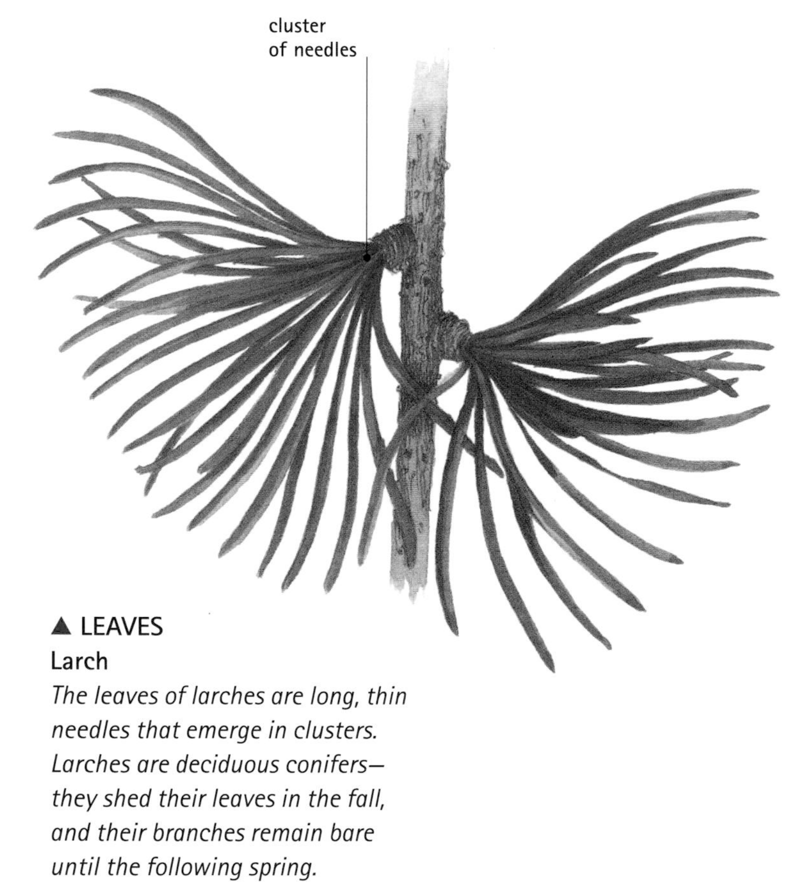

▲ LEAVES
Larch
The leaves of larches are long, thin needles that emerge in clusters. Larches are deciduous conifers—they shed their leaves in the fall, and their branches remain bare until the following spring.

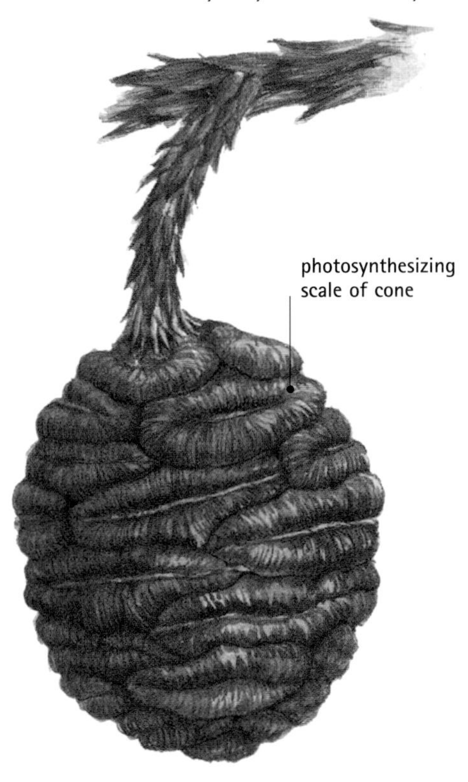

▼ SEED CONE
Giant sequoia
The cones, which contain the female reproductive organs and the seeds, hang down and grow from 1.5 to 2.5 inches (4–6 cm) long. They take two growing seasons to mature but may stay closed for 20 years.

▲ SEED CONE
Larch
In larches, the seed cones are up to 3.5 inches (9 cm) long. They grow erect on the branches and ripen to release their seeds within a year.

CLOSE-UP

Mycorrhizae

Conifers, along with many other plants, depend on a special association with fungi living in and around the tree roots. This intimate association is called a mycorrhiza, meaning "fungus root." Trees depend on their fungus to help them acquire water and nutrients from the soil. The fine threadlike fungal tissues, which are called hyphae, can explore much farther through the soil than the tree roots can alone. The hyphae obtain water and nutrients from the soil, and some of these are eventually absorbed by the tree from the hyphae. In exchange for these nutrients, the fungus takes some of the sugary products of photosynthesis from the tree, so both partners benefit from the association.

Most conifers have what are called ectomycorrhizae, in which a sheath of fungal hyphae grows around the outside of the smallest roots. This fungal sheath physically protects the root from drying out and also forms a protective layer against many disease-causing bacteria and fungi. The fungus also penetrates between the cortical cells (those near the root surface), forming a netlike structure inside the root. Roots with ectomycorrhizae become stubby and thickened, and look different from plants with roots that do not have a fungal association.

Conifer leaves are good at resisting extreme weather. Their long, thin shape gives them a small surface area. In areas where water is limited, a small leaf area helps conserve water —a flatter leaf would lose too much water by evaporation. Conifer leaves are covered by a thick, waxy cuticle that also reduces evaporation. Needlelike leaves and the poise of the branches also help shed excess snow, the weight of which would otherwise damage the tree.

Conifer roots

Roots have two key functions: to anchor the tree in the soil and to extract water and soil nutrients. The roots of conifers tend to be shallow but spread over a wide area. Many conifers can take root in cracks in almost bare rock and are capable of growing on the steepest of slopes, where they help stabilize loose, stony soil. The roots of conifers are also capable of extracting essential water from soil where it is in very short supply, either in very hot, dry soils, or in frozen soil in which the water is locked up as ice.

Like all other living parts of the plant, roots also need oxygen for respiration, and oxygen is lacking in waterlogged soil. The North American swamp cypress produces knobby, woody "knees." These are aerial roots that point up out of the water and mud. They contain air channels that allow oxygen to move into the deeper roots under water.

Internal anatomy

CONNECTIONS

COMPARE giant sequoia wood with ***APPLE TREE*** wood. The wood from sequoia and other conifers does not contain large vessel elements but has tracheids instead. Apple tree wood contains both vessel elements and tracheids.

The anatomy of a conifer trunk enables it to carry out the functions of support, transport, and protection. The bark of sequoias is fibrous and very thick, often more than 1 foot (30 cm) deep. It is as effective as asbestos for fire protection and contains tannic acid, which was once widely used in fire extinguishers.

The bark protects the layer of living cells below it. This layer, called cambium, is where the new tissue (secondary growth) is created as the tree grows in girth, or width, to support the increasing weight of the tree.

On the outside of the cambium, there is a thin layer of phloem cells. The phloem tissue conducts the sugary products of photosynthesis from the leaves to the rest of the plant. Phloem cells are long and thin. They are called tube elements, because they have pitted areas on the sidewalls and end walls where they join other tube cells.

The bulk of the tree trunk is made up of xylem tissue. Xylem tissue conducts water and provides the tree's strength. In conifers, the xylem is composed of tracheids. These are long, narrow cells that are connected by pits, so water can move freely through and between them. Tracheids vary in size among different species of conifers. In the Utah juniper, they average just 0.05 inch (1.2 mm) long, whereas the Californian redwood has large tracheids averaging 0.3 inch (7.4 mm) long.

Conifer wood also has a small amount of packing tissue made of cells called parenchyma. Most of this tissue is laid down in rays that run perpendicular to the tracheids. These rays also contain ray tracheids that conduct water and other chemicals between the rings in the trunk. Therefore, conifers can reroute supplies if one area of the trunk is damaged.

COMPARATIVE ANATOMY

Soft and hard

The term *hardwood* is commonly used for the wood from broad-leaved angiosperm trees, such as apple and oak trees, whereas *softwood* is used for conifers. However, these terms can be misleading. One of the hardest woods in the world is from a conifer, yew, and therefore is classed as a softwood, whereas one of the softest woods, balsa, is from a tropical angiosperm and so is classed as a hardwood. The real difference between these two groups is in their structure. Softwoods contain only tracheids and so have a simpler structure than hardwoods. Hardwoods contain tracheids and vessel elements.

▼ CROSS SECTION OF TRUNK
Giant sequoia
Beneath the thick bark lies the pale, living sapwood, which consists of phloem, cambium, and some xylem. The heartwood is heavily thickened dead xylem that is stained deep reddish-brown by resin.

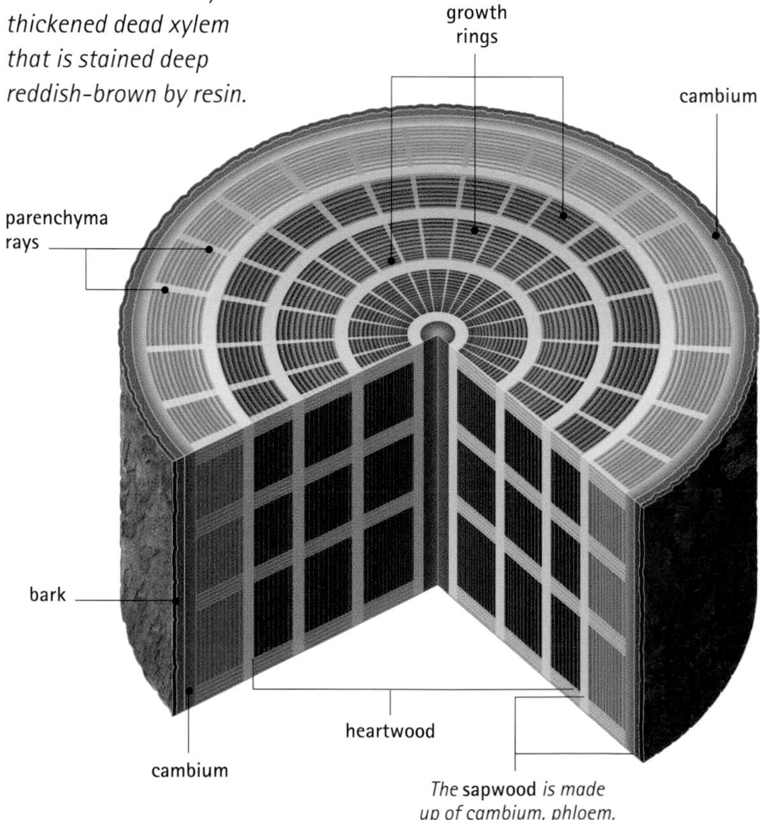

The sapwood *is made up of cambium, phloem, and xylem tissues.*

Resin ducts

Dotted among the densely packed tracheids are spaces called resin ducts in which resin accumulates and flows. Secretory cells line the resin ducts. There are a number of complex chemicals in resin. Some of these chemicals act as antifreeze, enabling conifers to survive extreme cold. Other chemicals help protect the wood from attack by bacteria, fungi, insects, and other pests. Resin exudes from wounds, drying to form a waterproof seal, and often coats leaves and stems, too, giving conifers their characteristic piney smell.

Tannins and other chemicals are laid down in the older, innermost sections of the trunk. In the California redwood and giant sequoia, these chemical deposits turn the heartwood a rich red, in contrast to the pale, living sapwood in the outermost parts of the trunk.

Conifer roots and leaves

The roots take up water and soil minerals and transport them to the rest of the plant. Most of the water uptake is through the youngest roots. Water enters through the cortex, a thick outer layer of cells, and passes through and between these cells until it reaches a band of tissue called the endodermis (inner skin). The endodermal cells have a waterproofed layer called the casparian strip, which helps control the movement of water and dissolved minerals into and out of the root core. A ring of phloem tissue surrounds the xylem core. As a root matures, it develops a corky layer called the peridermis (outer skin), equivalent to the bark of the trunk and branches.

In the leaves, the epidermis (the outer layer of cells) is covered with a thick waxy cuticle.

CLOSE-UP

Tree rings and dendrochronology

A cut across a tree trunk reveals a pattern of concentric rings in the wood. Tree rings form in seasonal climates, because the tree lays down wood at different rates according to the weather. In spring, when there is plenty of moisture available, the tree can grow quickly and makes cells with a large diameter (known as spring wood). As summer progresses, the tree grows more slowly and lays down cells with a narrower diameter (summer wood). Counting tree rings is a fairly reliable way to calculate a tree's age. The science of dendrochronology uses the variability in patterns of annual tree rings to examine the weather of centuries past. Wood from bristlecone pines has enabled scientists to build up a "weather diary" going back 8,200 years.

▼ RESIN DUCT
Giant sequoia
Resin ducts occur within the dense xylem tissue. Secretory cells line the ducts and produce a range of chemicals that protect the tree.

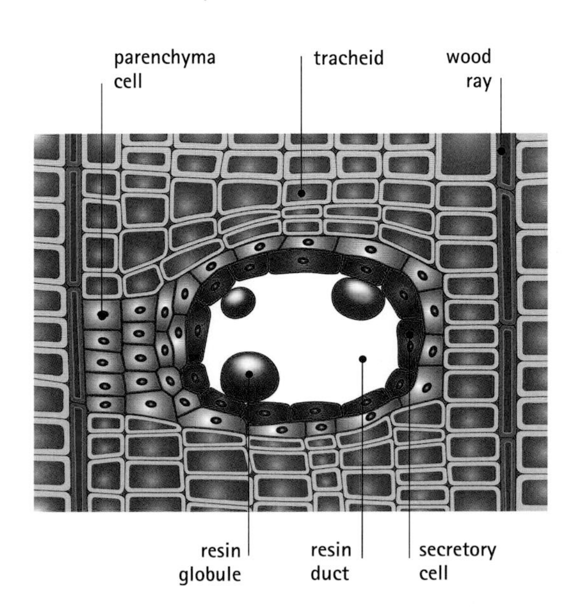

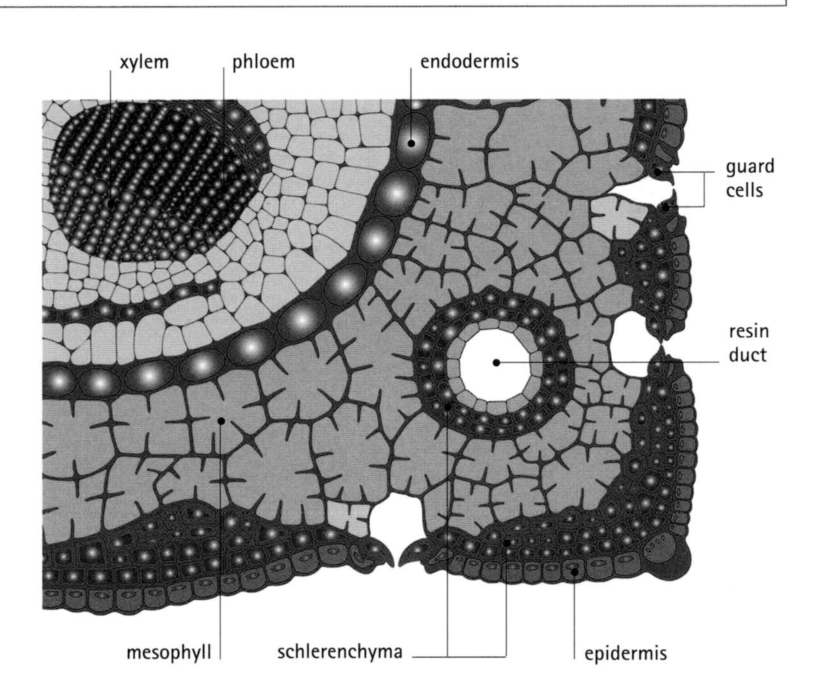

Dotted in the epidermis are holes called stomata, each surrounded by two guard cells. Many conifers have stomata in sunken pits, which protect them from drying winds.

The packing cells of a leaf, called mesophyll, are green because they contain thousands of tiny organelles ("mini-organs") called chloroplasts. Photosynthesis takes place in the chloroplasts.

Inside the leaf, one or several vascular bundles are visible as leaf veins. They contain the xylem and phloem tissues that conduct water, nutrients, and the products of photosynthesis between the leaf and other parts of the plant. Most species of conifers also have resin ducts in their leaves.

▲ LEAF CROSS SECTION
Giant sequoia
The main body of the leaf is composed of cells called mesophyll, which contain chloroplasts. The leaf has several vascular bundles running through it. Epidermal cells encase the leaf, and dotted within these cells are openings called stomata.

Reproductive system

CONNECTIONS

COMPARE pollination in the giant sequoia with pollination in ***ORCHIDS***. Orchids rely on insects to move the pollen from one plant to another. They produce small amounts of pollen in sticky bundles. The sequoia and other conifers are wind-pollinated. These trees release masses of dry pollen into the air.

Unlike angiosperms, which produce flowers, gymnosperm plants produce two types of cones. The male cones produce pollen, whereas the female cones have ovules that develop into seeds after fertilization. Almost all conifers are monoecious, producing both male and female cones on the same tree.

In the giant sequoia, male cones cover the outer branches in winter and produce clouds of golden pollen. The pollen is produced from two structures called microsporangia that lie on the underside of each cone scale. Each pollen grain has air sacs that increase its buoyancy, enabling it to drift in the slightest breeze.

Pollination and fertilization

At pollination, the female cones are tiny, about the size of a grain of wheat. The ovules are usually on the upper surfaces or edges of the

GENETICS

DNA from "dad"

California redwood has a very unusual inheritance system for its mitochondrial DNA. Mitochondria are the energy-producing organelles (mini-organs) inside cells. They have their own DNA, which is separate from the DNA in the cell's nucleus and is inherited from only one parent. In almost every organism studied, the mitochondrial DNA (mtDNA) is from the maternal line. In the case of plants, therefore, the mtDNA comes from the ovule rather than from the pollen. In California redwood, however, the mtDNA is paternally inherited, coming from the male sex cell (pollen).

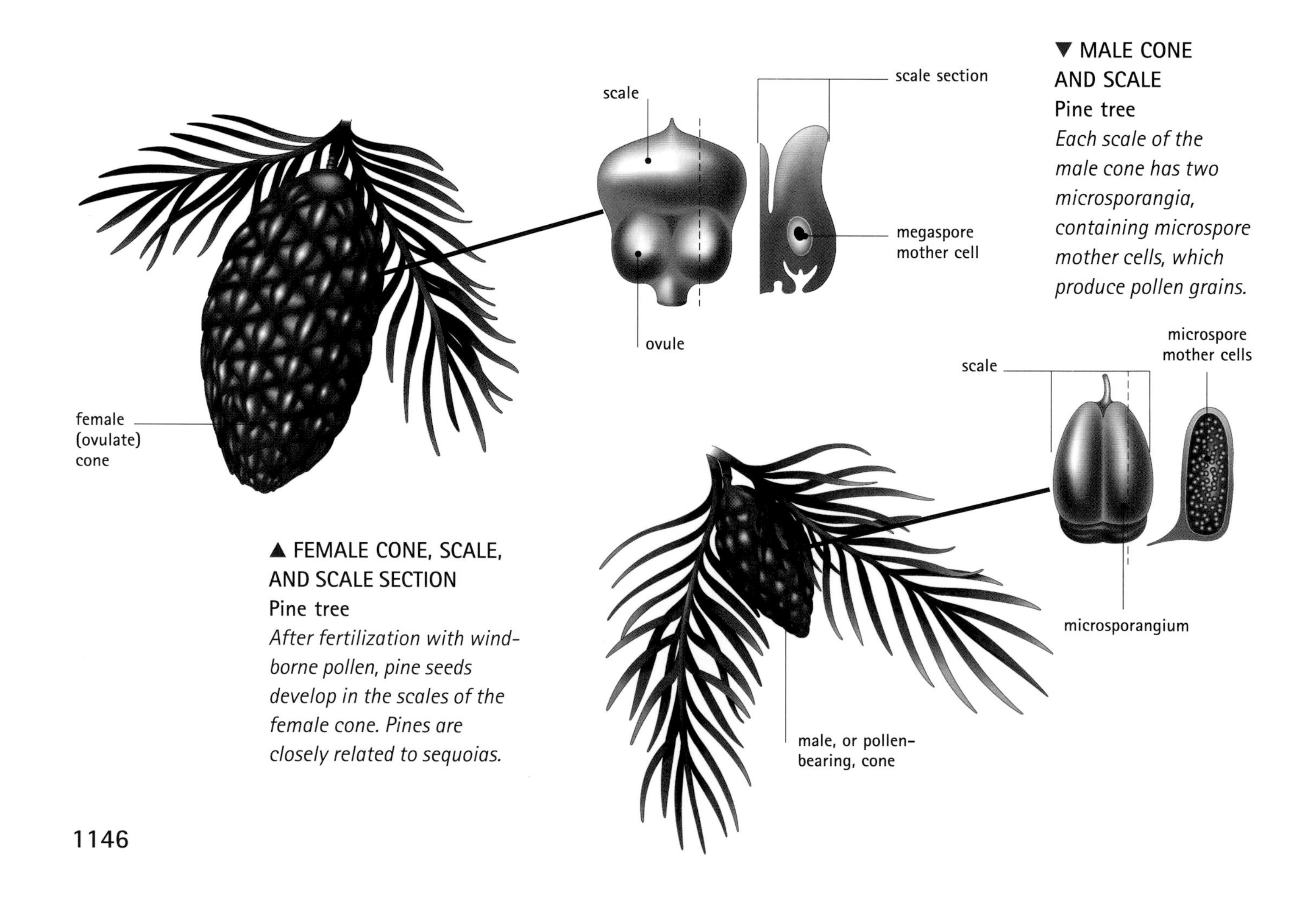

▼ MALE CONE AND SCALE
Pine tree
Each scale of the male cone has two microsporangia, containing microspore mother cells, which produce pollen grains.

▲ FEMALE CONE, SCALE, AND SCALE SECTION
Pine tree
After fertilization with wind-borne pollen, pine seeds develop in the scales of the female cone. Pines are closely related to sequoias.

◀ **SEED CONES**
Pine tree
These cones clearly show the spiral arrangement of the scales. One set of spirals runs steeply around the cone in one direction, and the other set runs at a more shallow angle in the opposite direction.

cone scales. The ovule catches the free-floating pollen grains on a sticky droplet exuded from a small pore, called a micropyle, in the outermost "skin" (integument) of the ovule.

The pollen is pulled into the ovule as the sticky liquid dries or is absorbed back into the ovule. The pollen grain then grows an extension called a pollen tube, down which the nucleus moves until it can fuse with the egg nucleus in the ovary.

Seed cones

Once the ovules have been fertilized, the scales of female cones grow, produce chlorophyll, and turn green. In the giant sequoia, the cones grow quickly and by the end of the first summer are usually more than three-quarters their full size. By the end of the second growing season, when they are aged 18 to 20 months, the cones have grown to between 1½ and 2½ inches (4–7 cm) long. They are then mature, and their seeds are usually viable and capable of germinating. When mature, most conifer cones become woody and dry and open to release their seeds.

CLOSE-UP

Cone patterns

The scales on a pine cone are in two sets of spirals. One set runs steeply up the cone, while the other can be traced in the opposite direction at a shallower angle. A mathematical phenomenon is hidden in these spirals. If you count the number of scales in any of the steep spirals and in any of the shallow ones, the two numbers should fall on the Fibonacci series (1, 1, 2, 3, 5, 8, 13, 21 . . . in which each number after the first two is the sum of the two before it). This property is a result of the anatomy of the developing cone. As the cells that form the next cone scale, called primordia, begin to bulge, they grow at an angle of 137.5 degrees from the scale developing behind it. This angle occurs naturally because it provides the most space for growth. The angle creates the Fibonacci pattern, which can also be seen elsewhere in the natural world, from pine cones to sunflowers and pineapples.

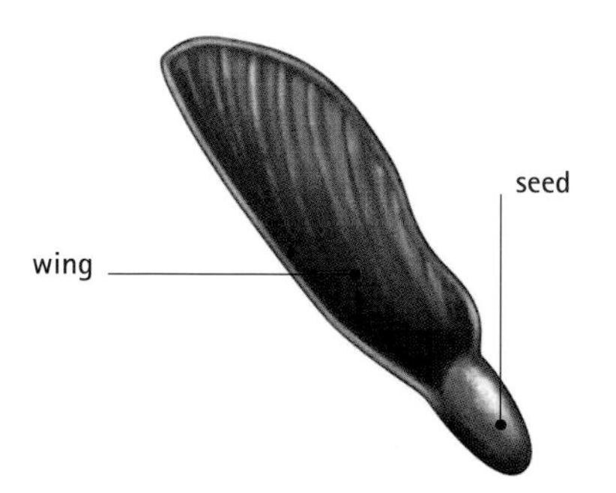

◀ **WINGED SEED**
Giant sequoia
The winged seeds, which are under 0.1 inch (5 mm) long, are usually dispersed by the wind. One cone may produce up to 300 seeds, although the average is 230.

Giant sequoias, however, behave differently because most of its cones stay green and tightly closed and can remain like this on the tree for 20 years or more.

Seed dispersal

Most conifer seeds are dispersed by wind. Seeds usually have a single wing, which helps to carry them away from the parent tree. In the giant sequoia, each cone may produce up to 300 seeds. The seeds are less than 0.1 inch (5 mm) long. It is estimated that a mature tree produces 300,000 to 400,000 seeds each year. These seeds are released from treetops or dispersed by animals, but release can occur only when the cone has turned brown. Some sequoia seeds are shed when the cone scales dry out in hot weather in late summer, but most are not released until the cones dry out from the heat of forest fires or from damage caused by insects.

▼ *The Douglas squirrel, or chickaree, feeds on the fleshy scales of green sequoia cones but discards the seeds. In this way, the squirrel helps disperse seeds to areas away from the parent tree.*

COMPARATIVE ANATOMY

Conifers with berries

Not all conifer cones are made of dry scales. The English yew and the plum yew, which comes from the tropical and temperate forests of China and Japan, produce berrylike cones. In conifers that produce berries, the female cones are highly reduced. Most conifers produce a female cone with lots of scales, but berry-producing species have just one or two scales and one ovule. As the ovule develops into a seed, the scale to which it is attached develops into a structure called an aril, which partly encloses the seed. The aril is brightly colored and fleshy, and looks like a berry (technically, true berries are formed only in angiosperms, in which the entire ovary wall becomes fleshy). Like true berries, arils are attractive to birds that eat the flesh and disperse the seeds.

Beetles are one of the few things that will turn sequoia cones brown. The larvae of the longhorn beetle *Phymatodes nitidus* live in the cone scales, chewing tunnels. If a beetle larva severs the vascular bundles (veins) that supply the cone, the tissue dries out and the scales shrink, allowing the seeds to fall out. Fire is also an agent of seed release. The hot air from a forest fire dries the cones and opens the scales, so huge numbers of seeds are released onto ground that has been freshly cleared. The Monterey pine and many other conifers of fire-prone regions have what are known as serotinous cones—cones that open with fire.

The chickaree, or Douglas squirrel, is an agent of seed distribution. It feeds on the fleshy scales of younger cones, dislodging and discarding the seeds since they are too small to bother with. Other conifer seeds are eaten, and dispersed, by woodpeckers, scrub jays,

▲ *The hot air from forest fires can dry sequoia cones so they release their seeds. The thick bark of the giant sequoia contains tannic acid, which can protect the tree from the flames.*

ground squirrels, chipmunks, and other birds and mammals.

Germination and survival

Before sequoia groves were managed by humans, forest fires, mostly caused by lightning, occurred every three to eight years. These fires would clear large areas of the forest floor, allowing sequoia seeds to germinate and grow. The seeds need bare soil, moisture, and plenty of sunlight to germinate, which can only happen naturally after a fire has burned away the duff (the partly decayed twigs and leaves on the forest floor) and small competing plants and trees.

However, because fires in sequoia groves have been controlled by forest workers in the last 100 or so years, the forest floor is permanently covered by a dense understory, preventing the germination of sequoia seeds. In some groves, the youngest trees are more than 100 years old.

Unlike adult trees with their thick fire-resistant bark, sequoia seedlings and saplings are easily killed by forest fires. (Adult trees can survive after having 95 percent of their foliage burnt away.) In addition, many animals such as chipmunks, sparrows, and finches eat most of the seedlings. If a seedling does manage to survive, giant sequoias grow fast and produce cones after about 20 years. The most common cause of natural death in the giant sequoia is lightning, which can be powerful enough to split the massive trunk in two.

ERICA BOWER

FURTHER READING AND RESEARCH

Bell, P. R. and A. R. Hemsley. 2005. *Green Plants: Their Origin and Diversity*. Cambridge University Press: New York.

Cafferty, S. 2005. *Firefly Encyclopedia of Trees*. Firefly: Toronto.

Heywood, V. H. 2007. *Flowering Plant Families of the World*. Firefly: Toronto.

Index

Page numbers in **bold** refer to main articles; those in *italics* refer to picture captions, or to names that occur in family trees.

J

K

L

M

N

O

P

R

S